SPRINGER
LAB MANUAL

Springer
Berlin
Heidelberg
New York
Barcelona
Budapest
Hong Kong
London
Milan
Paris
Santa Clara
Singapore
Tokyo

Valdis Berzins (Ed.)

Basic Cloning Procedures

With 24 Figures

 Springer

PROFESSOR DR. VALDIS BERZINS
University of Latvia
Biomedical Research and Study Centre
Ratsupites Str. 1
LV 1067 Riga
Latvia

ISBN 978-3-642-48977-8   ISBN 978-3-642-71965-3 (eBook)
DOI 10.1007/978-3-642-71965-3

Library of Congress Cataloging-in-Publication Data
Basic cloning procedures / Valdis Berzins (ed.). p. cm. – (Springer lab manuals) Includes biblio-
graphical references and index. ISBN 978-3-642-48977-8   1. Molecular
cloning-Laboratory manuals. I. Berzins, Valdis, 1943 -. II. Series: Springer lab manual.
QH442.2.B37 1998 572.8'645-dc21

Production: PRO EDIT GmbH, Heidelberg
Cover design: design & production GmbH, Heidelberg
Typesetting: Mitterweger Werksatz GmbH, Plankstadt
SPIN 10543880   31/3137   5 4 3 2 1 0 - Printed on acid free paper

Preface

Experience in laboratory experiments in molecular biology is important to students, researchers and practioners in the biological and medical sciences. This book is designed to introduce the reader to basic methods used in the isolation, cloning and analysis of genetic material.

The text includes theory and fundamental methods in sufficient experimental detail which have been successfully used in our studies of viral genome and eukaryotic gene regulation. Specific protocols together with troubleshooting tips are given for RT-PCR amplification, gene cloning, hybridization analysis and sequencing of nucleic acids, PCR-based site-specific mutagenesis, analysis of protein DNA-specific interaction, cell-free protein synthesis and product electrophoretic and immunological analysis.

The methods chosen have been approved in several international Molecular Biology Courses and without doubt can be used as a manual in basic molecular studies of DNA and RNA.

The editor is grateful to all the contributors for their cooperation and would like to express his thanks to the publishers, especially Dr. J. Lindenborn from Springer-Verlag, for their cooperation, patience and assistance during the preparation of this book.

Riga, Latvia   V. BERZINS

Contents

Contents

cDNA Synthesis and Cloning

Juris Steinbergs[*1] and Aleksander Tsimanis

Introduction

The isolation of intact messenger RNA and its conversion into cDNA copies by avian or Moloney murine reverse transcriptase, as well as subsequent amplification of gene transcripts by the PCR technique, are becoming increasingly important tools in molecular biology. At present, these techniques have been often necessary and widely used for the analysis of individual mRNA levels in cells and tissues by Northern blot analysis, nuclease protection analysis and in situ hybridization. Another important application of RNA templates is the construction of representative cDNA libraries in order to clone genes, to investigate their molecular structure and to express them in prokaryotic and/or eukaryotic cells (Belyavsky 1989).

Procedures which are presented herein allow the isolation of mRNA with high purity and quality and the construction of a corresponding cDNA library within two or three working days.

Total RNA Isolation

A number of techniques are currently used for the isolation of RNA from tissues and eukaryotic cells and almost all of them use guanidine thiocyanate, the most widely used reagent for the isolation of RNA from cells rich in ribonuclease (Chirgwin 1979; Chomczynski and Sacchi 1987). There are two potential difficul-

Principle and Applications

[*] Corresponding author: Juris Steinbergs; phone: +371–2–428116; fax: +371–2–427521; e-mail: juris@biomed.lu.lv
[1] University of Latvia, Biomedical Research and Study Centre, Ratsupites Str. 1, LV1067 Riga, Latvia

ties that are encountered in the isolation of intact RNA – degradation by ribonucleases and contamination by high molecular weight DNA. Ribonuclease is a very stable enzyme which can be found inside and outside of cells. Therefore, the use of guanidine thiocyanate, a strong chaotropic denaturant which unfolds ribonucleases at a rate which exceeds the rate of RNA hydrolysis is needed from the moment of cell lysis or tissue sample homogenization. Very effective is a combination of guanidine thiocyanate with β-mercaptoethanol which reduces disulphide bonds of the enzyme and thus promotes the denaturation and, ultimately, destruction of ribonuclease activity. At least five companies (Clontech, GibcoBRL, Pharmacia, Qiagen and Stratagene) produce RNA isolation kits which provide a simple and fast method for the preparation of total RNA from bacteria, cultured cells and tissues. Other chemical compounds, such as sodium dodecyl sulphate (SDS) or N-laurylsarcosine and urea, denaturate ribonucleases, but the inactivation of their activity is not always effective enough.

In our lab we have used the guanidine thiocyanate/LiCl method described by Cathala et al. (1983), which is not designed as a commercial kit, to prepare total RNA from a variety of fresh and frozen sources (cells and tumours). This method combines the strong inhibition of ribonucleases with the selective precipitation of RNAs with lithium chloride and provides a good yield of RNA free from protein and DNA contamination.

Isolation of Poly(A)⁺mRNA

Usually poly(A)$^+$mRNA is prepared in two steps: first, the isolation of total RNA, followed by affinity purification on oligo(dT)-cellulose to select poly(A)$^+$mRNA, which constitutes 1–3 % of the total RNA population (Aviv and Leder 1972). Under high salt concentrations, polyA tails of mRNA will hybridize to the oligo(dT)$_{12-18}$ (usually short oligodT are covalently bound to insoluble support, such as cellulose and other magnetic and non-magnetic resins). Pure poly(A)$^+$mRNA is eluted in a small volume of water or low-salt buffer as a decrease in the salt concentration will disrupt mRNA-oligo(dT) complexes. Subsequently, poly(A)$^+$mRNA can be fractionated by sucrose density gradient centrifugation.

Generation of cDNA Library

The development of cDNA cloning techniques has been an increasingly powerful tool in molecular biology research and that is why considerable effort has been devoted to optimizing conditions for full-length cDNA clone preparation. Three main approaches have been used to generate cDNA libraries, and these have tried to preserve as much of the original sequences as possible in order to improve cloning efficiency, and to facilitate screening and subsequent analysis (Kimmel and Berger 1987). Originally cDNA cloning manipulations involved S1-nuclease digestion aimed at cleavage of the hairpin loop present in cDNA, which is used to prime the synthesis of the second strand catalysed by Klenow DNA polymerase (Efstratiadis 1976; Rougen 1976). Since the use of S1-nuclease leads to the loss of the 5'-terminal sequence of the mRNA, several methods have been described to avoid the use of S1 nuclease in the construction of cDNA libraries. The procedure described by Okayama and Berg (1982), as later modified by Gubler and Hoffman (1983), is currently the most widely used technique for cDNA cloning. In the first step, first-stranded cDNA is synthesized using the classical oligo(dT) priming of poly(A)$^+$mRNA in the presence of sodium pyrophosphate at an optimal concentration of 4 mM, which prevents the formation of nearly all hairpins. Next, the second-strand synthesis is performed using RNaseH to create nicks and gaps in the hybridized mRNA for the DNA synthesis and repair by DNA polymerase I and *E. coli* DNA ligase. It provides a rapid, highly efficient generation of a cDNA library and cloning of blunt end ds-cDNA into cloning vectors.

Reverse-Transcriptase PCR

Since its introduction 10 years ago, DNA amplification by the use of the polymerase chain reaction (PCR) has proven to be an extremely valuable tool for PCR-based cDNA library construction and the isolation of structurally or functionally related cDNAs, pre-characterized cDNAs, and cross-species cDNAs as well as cDNA with unknown sequences at the 5' end. This technique was shown to require only nanogram amounts of total RNA as starting material. The most critical parameter for generation of the representative cDNA library is the quality of the cDNA molecules used as template for the PCR amplification.

Therefore, priming of the poly(A)$^+$mRNA with sequence-specific or random primer is preferred over using the oligo(dT). The amplified cDNA fragment can then be cloned or directly sequenced.

Materials

Caution: The key to obtaining a good preparation of mRNA is to minimize ribonuclease activity during the initial stages of extraction and to avoid accidental introduction of trace amounts of ribonuclease from the glassware and solutions. Gloves are worn at all times. Laboratory glassware must be generally treated for 12 h at room temperature with a solution of 0.1% diethylpyrocarbonate, followed by heating to 100 °C for 15 min. Sterile, disposable plastic ware is essentially free of RNase and can be used for preparation and storage of solutions without pretreatment. All buffers (except for Tris-containing solutions) are made in diethylpyrocarbonate (DEPC) treated double glass-distilled water (0.1% DEPC, boiled for 30 min and then auto-claved for 1 h).

All reagents are molecular biology grade or of the highest purity available.

Equipment
- Thermocycler (Amplegen PCR3, LIAP)
- Microcentrifuge (Heraeaus Sepatech, Biofuge B)
- Vortex (TM Scientific Ind., Genie 2)

Reagents
- Oligo(dT) cellulose, Type 7 (Pharmacia, #27–5543–02)
- Guanidine isothiocyanate (Merck, #4167.0250)
- Sephadex G-50 DNA Grade F (Pharmacia, #17–0573–01)
- NICK Spin Columns, Sephadex G-50 DNA grade F (Pharmacia, #17–0862–01)
- The following oligonucleotide primer pairs were used:
  5'-GGCCAGTGGATAGACAGATGG – VH const
  5'-GTGATGCTGGTGGAGTCTGG – VH 5' end
  5'-ATGGTGGGAAGATGGATACAG – VL const
  5'-GACATTGTGATGACCCAAAAC – VL 5' end
- 2YT Broth, powder (Gibco BRL, #M02712B)
- LB liquid medium
  yeast extract – 5 g/l
  Bacto tryptone – 10 g/l
  NaCl – 8 g/l

pH 7.0 adjusted with 10 % NaOH
- LB/Amp/X-gal/IPTG plates
  500 ml sterile LB agar (~50 °C) 2.5 ml of ampicillin solution (100 µg/ml)
  1 ml X-gal solution, X-gal solution, 2 % (w/v) in DMSO or DMFA
  2.5 ml IPTG solution, IPTG solution, 100 mmol/l
- AMV reverse transcriptase, FPLCpure (Pharmacia, #27–0922–02)
- M-MuLV reverse transcriptase, supplied with 5× reaction buffer (Fermentas, #EP0351)
- Ribonuclease inhibitor (Fermentas, #EO0311)
- *Thermus aquaticus* (Taq) DNA polymerase (Fermentas, #0071)
- Synthetic Oligonucleotides 5'-end Labeling Kit (Fermentas, #KO911)
- DNA Cloning Kit (Fermentas, #K1311)
- T-Cloning Kit (Fermentas, #K1211)
- DNA Ligation Kit (Fermentas #K1411)
- T4 DNA polymerase (Fermentas, #EP0061)
- DNA Polymerase I, Klenow Fragment (Fermentas, #EP0051)
- S1 nuclease (Fermentas, #EN0321)
- Ribonuclease H (Fermentas, #EN0201)

Buffers

- Lysis buffer
  5 M guanidine isothiocyanate
  50 mM Tris-HCl, pH 7.5
  10 mM EDTA
  8 % β-mercaptoethanol
- TS buffer
  10 mM Tris-HCl, pH 7.5
  0.5 % SDS
- TE buffer
  10 mM Tris-HCl, pH 7.5
  1 mM EDTA
- TBE buffer
  10 mM Tris-HCl, pH 8.3
  10 mM boric acid
  1 mM EDTA
- Loading buffer
  20 mM Tris-HCl, pH 7.6
  0.5 M NaCl

- Elution buffer
  10 mM Tris-HCl, pH 7.6
  1 mM EDTA
- Gel loading buffer
  2 % Ficoll
  0.5 % SDS
  50 mM EDTA
  0.2 % bromophenol blue
  20 % glycerol
- Extraction buffer
  50 mM Tris-HCl, pH 8.0
  0.2 % SDS
  0.3 M Na acetate
  4 mM EDTA

Buffered phenol is prepared by melting redistilled phenol at ~60 °C, adding 8-hydroxy-quinoline to 0.5 %, extracting five times with an equal volume of 1 M Tris-HCl, pH 8.0 and then with 0.1 M Tris-HCl, pH 8.0. The phenol is stored under 0.1 M Tris-HCl, pH 8.0, at 4 °C in a brown bottle.

- 10× PCR buffer (Fermentas):
  0.1 M Tris-HCl, pH 8.8 (at 25 °C)
  0.5 M KCl, 0.8 % Nonidet P40
- 10× T4 DNA polymerase buffer
  0.5 mM Tris-HCl, pH 8.5
  150 mM $(NH_4)SO_4$
  70 mM $MgCl_2$
  1 mM EDTA
  100 mM β-mercaptoethanol
- 10× ligation buffer
  0.66 M Tris-HCl, pH 7.6
  50 mM $MgCl_2$
  50 mM DTT
  10 mM ATP

1.1
Isolation of mRNA

Procedure

Extraction of Total RNA

1. Collect 5×10^7 hybridoma cells producing mouse monoclonal antibody S-26 (IgG1, κ) against preS2 antigen of Hepatitis B virus in 1.5 ml centrifuge tube and remove supernatant.

2. Resuspend cell pellet in 300 μl of lysis buffer. Lyse cells immediately! To facilitate lysis, homogenize the suspension by vortexing and divide into 2×150 μl.

3. Remove cell debris by centrifugation at 11 000 rpm for 15 min.

4. Transfer supernatant to a clean 1.5 ml tube.

5. Add 7 vol (1050 μl) of cold 4 M LiCl to the supernatant, mix (vortex) and then leave for 18 h at 4 °C.

6. Centrifuge at 11 000 rpm for 30 min. Discard supernatant.

7. Resuspend pellet in 4 vol (600 μl) of cold 4 M LiCl and then leave for 4 h at 4 °C.

8. Repeat step 6.

9. Dissolve RNA pellet in 200 μl of TS buffer with repeated vortexing and pipetting.

10. Heat at 100 °C for 3 min and cool on ice.

11. Add 200 μl of phenol saturated with TE buffer, vortex 15 sec and centrifuge at 11 000 rpm for 5 min.

12. Transfer aqueous layer to a fresh 1.5 ml tube and repeat deproteinization step.

13. Extract aqueous phase with chloroform:isoamyl alcohol (24:1).

14. Transfer upper (aqueous) layer to a fresh 1.5 ml tube.

15. Add 8 μl of 5 M NaCl and 500 μl of cold ethanol, mix and precipitate overnight at −20 °C.

16. Centrifuge at 12 000 rpm for 1 h at 4 °C. RNA will be in the pellet.

17. Wash pellet with 500 µl of 80 % cold (−20 °C) ethanol and centrifuge at 11 000 rpm for 15 min.

18. Dissolve RNA pellet in 50 µl of sterile 10 mM Tris-HCl, pH 7.5 and measure OD at 230, 260 and 280 nm by adding 2 µl to 198 µl TE buffer in a 200 µl quartz cuvette. 1 OD/ml corresponds to 42 µg RNA/ml.

19. Analyse 1–2 µg RNA by electrophoresis on a 1.5 % agarose gel in TBE buffer.

Note: When RNA is converted to cDNA with AMW reverse transcriptase followed by PCR reaction, total RNA can be used without additional purification.

Purification of poly(A)$^+$mRNA

1. Prepare an oligo(dT)cellulose column in a 1.0 ml glass column plugged with sterile glass wool (up to 3 mg of poly(A)$^+$ RNA can be processed per g of the oligo(dT)cellulose, Type 7).

2. Wash column with 5 vol of sterile loading buffer.

3. Dissolve RNA pellet in 100 µl 20 mM Tris-HCl, pH 7.5; heat RNA sample to 65 °C for 5 min, then cool on ice, add NaCl to 0.5 M and apply to the column.

4. Collect the eluate, heat to 65 °C, cool, and reapply it to the column.

5. Wash column with 5–10 vol of loading buffer until OD at 260 nm is equal to 0.

6. Wash column with 4 vol of loading buffer, containing 0.1 M NaCl.

7. Elute poly(A)$^+$RNA with 2–3 vol of sterile elution buffer.

8. Combine the RNA-containing fractions and repeat all steps to increase purity of the poly(A)$^+$RNA.

9. The yield from 10^8 cells should be 1–5 µg poly(A)$^+$RNA.

1.2
cDNA Synthesis

Procedure

Synthesis of the First cDNA Strand

Poly(A)$^+$mRNA is reverse-transcribed with the aid of avian mye-loblastosis virus (AMV RT) or Moloney murine reverse transcriptase (M-MuLV RT) using oligo(dT) or a specific oligonucleotide as the primer.

1. Add 5 µl of total RNA (1 µg/µl), 1 µl 2 M KCl, 5 µl of oligo(dT)$_{12-18}$ (0.5 µg/µl) or specific oligonucleotide (0.1 µg/µl) into a 0.5 ml microcentrifuge tube, heat to 70 °C for 5 min and cool slowly to 42 °C.

2. To the reaction mixture add:
   1 M Tris-HCl, pH 8.3, 0.2 M MgCl$_2$ 1.0 µl
   0.2 M dithiothreitol 3.0 µl
   10 mM dNTP mix 3.0 µl
   RNasin (40 U/µl) 1.0 µl
   40 mM Na pyrophosphate 1.0 µl
   dd H$_2$O 9.0 µl
   Final volume, 28 µl.

3. Mix gently and add 1 µl of AMV reverse transcriptase (18 U/µl), mix and incubate at 42 °C for 40 min.

4. To stop the reaction, incubate the tube at 100 °C for 3 min, when the second cDNA strand synthesis is performed by thermostable DNA polymerase.
   If the second cDNA strand synthesis is made by another method (see below) the dNTP mix usually contains [α-^{32}P] or [^{3}H] dNTPs, and the procedure is as follows:

5. Add 30 µl of phenol-chloroform (1:1) to the first strand reaction mixture and mix (vortex).

6. Prepare a Sephadex G-50 column in a disposable glass pipette plugged with sterile glass wool. Wash the column with several volumes of TE. For rapid purification of labelled DNA from unincorporated nucleotides we also recommend the use of Pharmacia NICK Spin Columns (#17–0862–01).

7. Apply the aqueous phase to the column. Wash the tube with approximately 50 µl of TE and load the washing on to the column. Connect the TE reservoir to the column so that the flow rate is about 0.5 ml/min.

8. Collect 10–12 fractions (0.3 ml) into microcentrifuge tubes. Measure radioactivity of each of the tubes by Cerenkov counting in a liquid scintillation counter. The leading peak of radioactivity consists of nucleotides incorporated into DNA, while the trailing peak consists of unincorporated [α-^{32}P] or [^{3}H] dNTPs.

9. Pool the radioactive fractions of the leading peak, then add NaCl to 0.2 M and 2.5 vol of 95 % ethanol and mix. Chill on dry ice for 1 h or at −20 °C overnight. Spin for 15 min at 4 °C in a microcentrifuge.

10. Remove supernatant with a micropipetting device and immediately add 1 ml of 75 % ethanol without agitation. Spin down for 1 min in a microcentrifuge and carefully remove the supernatant. Dry the pellet briefly under vacuum and dissolve in 40 µl of TE.

Synthesis of the Second cDNA Strand

Today, three methods are generally used:

- The synthesis of the second cDNA strand by thermostable DNA polymerase from *Thermus aquaticus* or *Thermus thermophilus* in a polymerase chain reaction.

- The synthesis of the second cDNA strand by *E. coli* DNA polymerase I or reverse transcriptase with subsequent cleavage of the hairpin loop with S1 nuclease.

- The synthesis of the second cDNA strand by *E. coli* DNA polymerase I in the presence of RNase H and T4 DNA ligase.

Synthesis of the Second cDNA Strand by Thermostable DNA Polymerase

PCR is a powerful method for the enrichment of specific target sequences. Both double- and single-strand DNA can serve as template for thermostable DNA polymerase. However, the dependence of the PCR on a pair of specific primers flanking the region of interest has generally limited its use to targets whose sequence is known beforehand.

1. Add in 0.5 ml microcentrifuge tube:
   mRNA-cDNA 5.0 µl (add last)
   10× PCR buffer 5.0 µl
   25 mM MgCl$_2$ 4.0 µl
   primer 1 (50 µg/µl) 2.0 µl
   primer 2 (50 µg/µl) 2.0 µl
   2.5 mM dNTP mix 2.0 µl
   BSA (2 mg/ml) 2.5 µl
   dd H$_2$O 26.5 µl
   Taq DNA pol 1.0 µl
   Final volume, 50 µl.

2. Load the program:
   1 min denaturing at 94 °C
   1 min annealing at 65 °C (usually 5 °C below the melting point of the primers)
   1 min extension at 72 °C, final extension – 5 min
   30 cycles.

3. Spin the tube briefly, add to the tube 30 µl of light mineral oil to prevent evaporation, put tube into thermocycler and start the program.

4. After 30 cycles, remove 5 µl of reaction mixture for electro-phoretic analysis in 1.5 % agarose gel (TBE buffer).

5. Add to the tube 5 µl of 10× T4 DNA polymerase buffer and 0.2 U T4 DNA polymerase. Incubate the tube at 37 °C for 15 min.

6. Extract the reaction mixture with 50 µl of chloroform, remove the aqueous phase and add to it 5 µl 3 M Na acetate, 150 µl of 95 % ethanol, mix well. Chill on dry ice for 1 h. Spin for 15 min at 11 000 rpm (4 °C) in a microcentrifuge, remove supernatant and wash pellet once with 200 µl of 75 % ethanol. Dry briefly and dissolve the pellet in 25 µl of TE.

DNA Polymerase I, Nuclease S1-Mediated Second cDNA Strand Synthesis

1. Add to 50 µl of the cDNA-RNA hybrid (0.1 µg/µl) 5.1 µl of 5 M NaOH and incubate for 40 min at room temperature.

2. Add 8.5 µl of 3 M sodium acetate (pH 4.5), mix well, then add 30 µl of H_2O and 225 µl of 95 % ethanol. Chill on dry ice for 1 h or at −20 °C overnight.

3. Collect DNA pellet by centrifugation as described above. Dissolve the dry pellet in 50 µl of TE.

4. Add into 1.5 ml microcentrifuge tube:
cDNA (50 ng/µl) 7.5 µl
0.5 M KPO_4, pH 7.4 10.0 µl
0.1 M $MgCl_2$ 5.0 µl
40 mM β-mercaptoethanol 2.5 µl
2.5 mM dNTP mix 1.0 µl
dd H_2O 64.0 µl

5. Mix gently, then add 10 µl of *E coli* DNA polymerase I Klenow fragment (1 U/µl), mix and incubate at 37 °C for 30 min.

6. Extract the reaction mixture with 100 µl of phenol-chloroform (1:1) and apply the aqueous phase to a G-50 column (see above). Collect fractions and measure radioactivity. Pool the leading peak fractions (approximate volume, 1 ml) and add:
5 M NaCl 62.0 µl
0.5 M Na acetate, pH 4.5 62.0 µl
0.3 M $ZnSO_4$ 10.0 µl
S1 Nuclease (1 U/µl) 3.0 µl
Mix gently and incubate at 37 °C for 30 min.

7. Extract the reaction mixture with 1.0 ml of phenol-chloroform (1:1), add 40 µl of 5 M NaCl and 2.5 ml of 95 % ethanol to the aqueous phase and precipitate at −20 °C overnight.

8. Spin for 40 min at 5000 rpm in a swing-out rotor. Remove supernatant and immediately add 3 ml of 75 % ethanol without agitation. Spin for 15 min at 5000 rpm, remove supernatant, dry the pellet completely in vacuum and dissolve in 100 µl of TE.

DNA Polymerase I, RNase H-Mediated Second cDNA Strand Synthesis

During second strand synthesis, up to 500 ng of single-stranded cDNA (i.e., 1 µg of RNA-DNA hybrid) can be processed in 100 µl.

1. Add into a 1.5 ml microcentrifuge tube:
   cDNA (50 ng/μl) 10.0 μl
   0.8 M Tris-HCl, pH 7.5, 0.2 M MgCl$_2$ 2.5 μl
   0.2 M (NH$_4$)$_2$SO$_4$ 5.0 μl
   2.0 M KCl 5.0 μl
   2.5 mM dNTP mix 1.5 μl
   BSA (1 mg/ml) 5.0 μl
   dd H$_2$O 65.0 μl

2. Mix gently and then add 1 μl RNase H (0.85 U/μl), 5 μl DNA polymerase I (5 U/μl), mix and incubate reaction mixture sequentially 60 min at 12 °C and 60 min at 22 °C.

3. To stop the reaction add 8 μl 0.25 M EDTA, pH 8.0.

4. Add 110 μl of phenol-chloroform and vortex. Spin for 3 min in a microcentrifuge. Collect the aqueous phase and repeat extraction.

5. Apply the aqueous phase to the G-50 Sephadex column (see above). Collect fractions and measure radioactivity. Pool the leading peak fractions and precipitate cDNA by adding NaCl to 0.2 M and 2.5 vol of 95 % ethanol. Precipitate at −20 °C overnight.

6. Spin for 15 min at 4 °C in microcentrifuge, wash once with 0.5 ml of 75 % ethanol, dry briefly and dissolve cDNA pellet in 50 μl of TE.

1.3
Cloning of Double-Stranded cDNA

A variety of methods have been used to link double-stranded cDNA to plasmid vectors. The small pUC plasmids are popular vectors for cloning of cDNAs as they have the following advantages:

- a large number of unique cleavage sites

- blue/white selection to distinguish between vector and recombinants
  Characterization of recombinants is also facilitated by

- the small size of vector plasmid, when restriction analysis is necessary

- the ability to determine the nucleotide sequence of DNA insert by direct sequencing of plasmid DNA
  The most commonly used procedures are:

- Cloning using blunt ends.

Double-stranded cDNA molecules are inserted into plasmid DNA that has been cleaved with a restriction enzyme, to generate blunt ends.

- Cloning using sticky ends.

1. Desired restriction sites to the termini of cloned double-stranded DNA are introduced by PCR using primers with restriction site in its 5' end. After cleavage with appropriate restriction enzyme, cDNA molecules are inserted into plasmid DNA that has been cleaved with a compatible enzyme (Jung 1990; Kaufman 1990).

2. Synthetic linkers are added to the termini of double-stranded cDNA. The remaining procedure is as in 1.

3. Complementary homopolymer tracts are added to double-stranded cDNA and to plasmid DNA. Vector and cDNA are then joined by hydrogen bonding between the complementary homopolymeric tails to form open, circular, and hybrid molecules. In this case ligation in vitro is not necessary.

The cloning results are improved, if cDNA fragments are isolated from polyacrylamide or agarose gel after PCR or after restriction.

Procedure

Cloning Using Blunt Ends

1. Add the following to the microcentrifuge tube:
   cleaved plasmid DNA (0.1 µg/µl) 1.0 µl
   cDNA (0.1–0.2 pmol) variable µl
   10× ligation buffer 1.0 µl
   dd H_2O variable µl
   T4 DNA ligase (1 U/µl) 1.0 µl
   Final volume, 10 µl.

2. Incubate overnight (6 h minimum) at 5 °C.

Cloning Using Sticky Ends

1. Add to the microcentrifuge tube:
   PCR products (50–100 ng/µl) 25.0 µl
   10× ligation buffer 4.0 µl
   T4 DNA ligase (1 U/µl) 2.0 µl
   T4 polynucleotide kinase (10 U/µl) 1.0 µl
   dd H_2O 8.0 µl
   Final volume, 40 µl.
   Incubate at 25 °C for 2 h at 5 °C overnight.

2. Heat-inactivate enzymes (e.g., 70 °C for 10 min), dilute to double volume in restriction enzyme buffer and cut with 5–10 U of restriction enzyme for 2 h at appropriate temperature.

3. Extract with 80 µl of chloroform. Remove the aqueous phase and add to it 20 µl of gel loading buffer.

4. Load the sample on a preparative 6 % polyacrylamide gel (acrylamide: bisacrylamide, 59:1) in TBE buffer.

5. After electrophoresis, stain gel with ethidium bromide (1 µg/ml) for 10 min and photograph under UV light.

6. Cut out pieces of the gel containing required DNA fragment.

7. Put gel pieces into 1.5 ml tube and squash them with a plastic micropipette tip.

8. Add 300–500 µl of extraction buffer and incubate 12–16 h with shaking at 37 °C.

9. Microcentrifuge the solution for 15 min, remove supernatant and extract with equal volume of chloroform. Remove aqueous phase and precipitate DNA with 2 vol of cold ethanol at −70 °C for 30 min.

10. Spin for 15 min at 4 °C in microcentrifuge, wash once with 0.5 ml 75 % ethanol, dry briefly and dissolve cDNA pellet in 10 µl of TE.

11. Check the recovery by electrophoresing 1 µl on a 1.5 % agarose gel in TBE buffer.

12. Add the following to a microcentrifuge tube:
    cleaved plasmid DNA (0.1 µg/µl) 1.0 µl
    cDNA (0.1–0.2 pmol) variable µl

10× ligation buffer 1.0 µl
dd H$_2$O variable µl
T4 DNA ligase (1 U/µl) 1.0 µl
Final volume, 10 µl.
Incubate overnight (6 h minimum) at 5 °C.

13. Heat-inactivate the enzyme and transform *E. coli*.

1.4
Transformation of *E. coli* Cells

The transformation of *E. coli* cells by the calcium chloride method yields approximately 2×10^7 transformants per µg of intact pBR322 or pUC 18 DNA with *E. coli* strains RRI and JM 109. *E. coli* strain JM109 has mutations providing a very convenient white/blue selection of transformants.

Procedure

Preparation of Competent Cells

All procedures with cells are performed under sterile conditions.

1. Prepare a starting culture by adding a freshly isolated single colony to 5 ml of sterile LB medium. Grow overnight at 37 °C.

2. Inoculate 0.3 ml of starter culture into 60 ml (use a sterile 750 ml shaker flask) of fresh LB medium and grow the cells with vigorous agitation at 37 °C to OD$_{540}$=0.2 (5×10^7 cells/ ml).

3. Chill the culture on ice for 5–7 min. Centrifuge the cell suspension at 3000 rpm in a sterile 100 ml glass centrifuge tube for 6–7 min at 4 °C.

4. Discard supernatant. Resuspend cells in half of the original culture volume (30 ml) of an ice-cold, sterile solution of 0.1 M CaCl$_2$.

5. Place cell suspension on ice for 15 min, then centrifuge at 3000 rpm for 5–6 min at 4 °C.

6. Discard supernatant. Resuspend cells in 1 ml of ice cold 0.1 M CaCl$_2$. Store the suspension on ice (or at 4 °C) for 24 h.

Transformation Procedure

1. To 0.1 ml of competent cells in a transformation tube add 10 µl of a DNA solution and incubate on ice for 30 min.

2. Heat-shock the cell-DNA mixture by placing the tube in a 42 °C water bath for 3 min and then 10 min at room temperature.

3. Add 1 ml of prewarmed (37 °C) LB medium to the suspension. Incubate at 37 °C with moderate agitation for 60 min.

4. Plate the suspension on selective, ampicillin-containing (usually 100 µg/ml) LB/Amp/X-gal/IPTG plates by spreading 0.05 ml aliquots (max 0.1 ml; in this case the plates should be dried well) per plate. Cells should be spread quickly, but gently. It would be wise not to spread all of the mixture at once; store, for example half of it, at 4 °C (cells can be also spread on filters).

5. Incubate the plates upside down at 37 °C overnight.

6. After using the required number of colonies for analysis, store the plates at 4 °C.

Growth of Clones for DNA Analysis

1. Prepare tubes with 3 ml of LB medium containing the antibiotic (ampicillin, 50 µg/ml).

2. Inoculate medium with a colony, using a loop.

3. Incubate tubes on a shaker overnight (150–200 rpm) at 37 °C.

4. Collect the cell suspension in 1.5 ml microcentrifuge tubes. Store the remaining suspension at 4 °C until analysis is completed.

Results

Quality and Yield of Total RNA

Figure 1.1A illustrates the ethidium bromide-stained agarose gel of the sample of total cellular RNA recovered from mouse hybridoma line S-26. The typical gel shows the discrete predo-

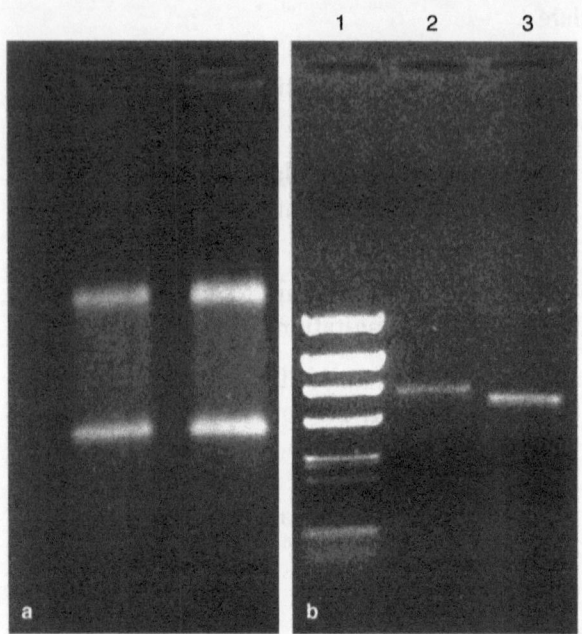

Fig. 1.a,b. Total cellular RNA isolated from the S-26 hybridoma cell line was separated on 1.5 % agarose gel and stained with ethidium bromide (**a**). Amplified cDNA from heavy and light chain variable regions encoding anti-preS2 mAb S-26 (**b**). *Lane 1* Marker fragments (pBR322/AluI); *lanes 2* and *3* amplification with γ and ϰ primers

minant 28S and 18S rRNA bands, moreover, the 28S RNA band stains with approximately twice the intensity of 18S RNA. Similar results were obtained with other B-cell (mouse hybridomas) and T-cell lines (EL-4 and Jurkat) and solid tumours (melanoma). Yields of total RNA from these different sources are around 5 μg per 10^7 cells or 100 mg of tissue. The quantity of RNA is calculated from the A_{260} readings, assuming that 1 A_{260} U per ml is equivalent to 42 μg of RNA per ml. Usually the total RNA isolated according to the established protocol has an A_{260}/A_{280} ratio of 1.7–2.0 indicating that contamination by proteins and polysaccharides is negligible. DNA is eliminated almost completely during precipitation of high molecular weight RNA with LiCl.

RT-PCR

There is no need to purify the total RNA into poly(A)$^+$mRNA for RT-PCR and that is why total RNA is used as template for reverse transcription reaction to generate cDNA and further to amplify a selected cDNA region. A discrete ~350 bp band for mouse Ig variable region light chain cDNA and ~380 bp band for Ig variable region heavy chain were obtained and demonstrated in Fig. 1.1B. Subsequently these fragments can be cloned into pUC plasmids and sequenced. We have experienced difficulty in cloning PCR-amplified DNA fragments. Because of a terminal transferase activity of Taq DNA polymerase (Clark 1988), additional treatment of PCR products with Klenow fragment, T4 DNA polymerase is usually necessary (Stoker 1990). At present, a fast and efficient one-step T-Cloning Kit is very popular (Fermentas, #K1211) that allows the cloning of PCR-derived DNA fragments without any additional end-polishing.

Taq DNA polymerase remains bound to the DNA even after deproteinization of the DNA fragment with phenol and therefore inhibits restriction endonuclease activity. Incorporation of proteinase K treatment of PCR product generated using primers containing terminal restriction sites prior to cutting with the corresponding enzymes significantly increase cloning efficiency (Crowe 1991).

Construction of cDNA Library

cDNA cloning of the Ig variable regions can be also performed by constructing and subsequent screening of a cDNA library. In this case poly(A)$^+$mRNA must be isolated. Usually the content of poly(A)$^+$mRNA is detected by its ability to incorporate dTMP in a reverse transcription reaction and the integrity of obtained mRNA preparation is verified by its ability to be efficiently transcribed into cDNA. Intact and pure poly(A)$^+$mRNA serves as an efficient template for cDNA synthesis. cDNA synthesis can be performed with random hexamer primers, oligo(dT) or sequence-specific primers. Anchored oligo(dT) primers bind at the junction of the polyA tails and therefore the probability of full-length cDNA transcripts is increased.

At an optimal incubation temperature (42 °C for AMV RT) the enzyme is capable of reading efficiently through secondary structure, but 4 mM sodium pyrophosphate inhibits the natural

RNAse H activity of AMV RT. Higher concentrations of this compound (6 mM) resulted in a strong inhibition of the first-strand cDNA synthesis (Rhyner 1986). The synthesized cDNA is very stable and ready also for PCR amplification.

Troubleshooting

The most important factor for successful synthesis of genes and generation of a cDNA library using the method described here is the quality of the mRNA template. The quality of RNA preparation means *purity* and *integrity*.

- The purity of total RNA is determined from the A_{260}/A_{280} readings. If the ratio is low, a significant amount of protein is present in the sample and an additional phenol/chloroform treatment is necessary.

- The second parameter is the A_{260}/A_{230} ratio; if this value is lower than two, the RNA sample is contaminated with guanidine isothiocyanate and we recommend to reprecipitate the RNA with ethanol.

- It is relatively simple to analyze an RNA sample on content of DNA by agarose gel electrophoresis. If a smear is present, the homogenization of cells or tissues is not efficient (high molecular DNA did not shear to small fragments). Next time, this stage must be performed properly.

- The integrity of total RNA is checked by agarose gel electrophoresis and ethidium bromide staining. The absence of two sharp bands on the stained gel: the 28S and 18S eukaryotic ribosomal RNAs and the 23S and 18S bacterial ribosomal RNAs, respectively, indicates that the RNA preparation is degraded. The simple causes of RNA degradation are (a) RNase contamination of buffers; (b) cells and tissues used for RNA isolation had not been properly stored; (c) RNase had been introduced during agarose gel electrophoresis.

- The presence of SDS, EDTA and high salt concentration affects the activity of reverse transcriptase and efficiency of cDNA synthesis.

- When there is a failure to clone PCR-derived DNA fragments check if corresponding oligonucleotide primers or synthesized DNA fragments were phosphorylated.

References

Aviv H, Leder P (1972) Purification of biologically active globin messenger RNA by chromatography on oligothymidylic acid-cellulose. Proc Natl Acad Sci USA 69:1408–1412

Belyavsky A, Vinogradova T, Rajewsky K (1989) PCR-based cDNA library construction: general cDNA libraries at the level of a few cells. Nucl Acids Res 17:2919–2932

Cathala G, Savouret JF, Mendez B, West BL, Karin M, Martail JA, Baxter JD (1983) A method for isolation of intact, translationally active ribonucleic acid. DNA 2:329–335

Chirgwin JM, Przybyla AE, MacDonald RJ, Rutter WJ (1979) Isolation of biologically active ribonucleic acid from sources enriched in ribonuclease. Biochemistry 18:5294–5299

Chomczynski P, Sacchi N (1987) Single-step method of RNA isolation by acid guanidinium thiocyanate-phenol-chloroform extraction. Anal Biochem 162:156–159

Clark JM (1988) Novel non-template nucleotide addition reactions catalyzed by prokaryotic and eucaryotic DNA polymerases. Nucl Acids Res 16:9677–9686

Crowe JS, Cooper HJ, Smith MA, Sims MJ, Parker D, Gewert D (1991) Improved cloning efficiency of polymerase chain reaction (PCR) products after proteinase K digestion. Nucl Acids Res 19:184

Efstratiadis F, Kafatos FC, Maxam AM, Maniatis T (1976).Enzymatic in vitro synthesis of globin genes. Cell 7:279–288

Gubler U, Hoffman BJ (1983) A simple and very efficient method for generating cDNA libraries. Gene 25:263–269

Jung V, Pestka SB, Pestka S (1990) Efficient cloning of PCR generated DNA containing terminal restriction endonuclease recognition sites. Nucl Acids Res 18:6156

Kaufman DL, Evans GA (1990) Restriction endonuclease cleavage at the termini of PCR Products. BioTechniques 9:304–306

Kimmel AR, Berger SL (1987) Guide to molecular cloning techniques. In Berger SL, Kimmel AR (eds), Methods in Enzymology, vol 152. Academic Press, New York, pp 307–316

Okayama H, Berg P (1982) High-efficiency cloning of full-length cDNA. Mol Cell Biol 2:161–170

Rhyner TA, Biquot NF, Berrard S, Borbely AA, Mallet J (1986) An efficient approach for the selective isolation of specific transcripts from complex brain mRNA populations. J Neurosci Res 16:167–181

Rougen F, Mach B (1976) Cloning and amplification of rabbit α- and β-globin gene sequences into E. coli plasmids. J Biol Chem 252:2209–2217

Sardelli AD (1991) Cloning of PCR products. Amplifications, 6:10–11.

Stoker AW (1990) Cloning of PCR products after defined cohesive termini are created with T4 DNA polymerase. Nucl Acids Res 18:4290–4291

Chapter 2

DNA Sequencing

ERIKS JANKEVICS[1]

Introduction

The chemical degradation method (Maxam and Gilbert 1977) and the dideoxy termination method (Sanger et al. 1977) are the basic techniques for sequencing DNA. Both of these methods depend on analytical polyacrylamide gel electrophoresis to resolve oligonucleotides with one identical end and one end varying in length by a single nucleotide. The enzymatic method has been improved over recent years. The Sanger method was developed as a basic method for large-scale DNA sequencing projects, using double-stranded DNA templates and the T7 DNA polymerase as a sequencing enzyme. A number of new modified enzymatic sequencing methods were developed recently: cycle sequencing, solid-phase sequencing, automated sequencing. In contrast, the original Maxam-Gilbert method has not been widely used for sequencing long DNA fragments.

Principle and Applications

The chain termination method involves the synthesis of a DNA strand by DNA polymerase in vitro using a single-stranded or double-stranded DNA template. Synthesis is initiated at only one site where the oligonucleotide primer anneals to the template. The synthesis reaction is terminated by the incorporation of a nucleotide analog that will not support continued DNA elongation, hence the name chain termination. The chain terminating nucleotide analogs are 2',3'-dideoxynucleotide 5'-triphosphates (ddNTPs); these lack the 3'-OH group necessary for DNA chain elongation. When proper mixtures of dNTPs and one of the four

[1] University of Latvia, Biomedical Research and Study Centre,
Ratsupites Str. 1, LV1067 Riga, Latvia
phone: +371–2–428116; fax: +371–2–427521;
e-mail: jankevic@hotmail.com

ddNTPs are used, enzyme-catalyzed polymerization will be terminated in a fraction of the population of chains at each site where the ddNTP can be incorporated. Four separate reactions, each with a different ddNTP, give complete information. A radioactively labeled nucleotide is also included in the synthesis, so that the labeled chains of various lengths can be visualized by autoradiography after separation by high-resolution electrophoresis.

In the original procedure, primer extension was catalyzed by the Klenow fragment of *E. coli* DNA polymerase I. T7 DNA polymerase (Sequenase) does, however, have significant advantages over the Klenow fragment for sequencing:

- Because of its processivity and high rate of polymerization, longer chain-terminated fragments (>1 kb in length) can be generated very rapidly, with a more even distribution of label between fragments. This allows a greater length of sequence to be determined reliably from a single set of sequencing reactions.

- Because of its tolerance for substrate analogs, the same set of sequencing mixes may be used with either ^{32}P or ^{35}S. In contrast, the Klenow fragment requires separate mixes for the two labels.

The major practical difference when using T7 DNA polymerase rather than the Klenow fragment is that the primer extension reactions are performed in two stages, a 'labeling' reaction and a 'termination' reaction. The two stages are required because the enzyme uses dideoxynucleotides very readily. To permit synthesis of long-chain terminated fragments, dideoxynucleotides are therefore excluded during the first stage, then added for the second. Even so, the total time required for these reactions is significantly less than the time required for reactions with the Klenow fragment.

This protocol includes all of the DNA sequencing steps: DNA transfection or transformation, isolation, sequencing and electrophoresis. Both single-stranded (Sect. 2.1) and double-stranded (Sect. 2.1) DNA template preparations are described. The principal scheme of double-stranded DNA sequencing using T7 polymerase is shown in Fig. 2.1.

1.Denaturation

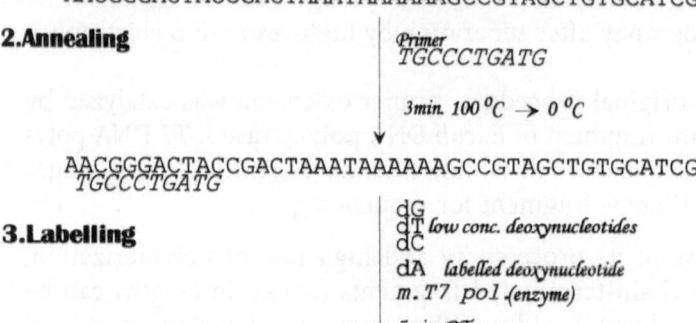

```
AACGGGACTACCGACTAAATAAAAAAGCCGTAGCTGTGCATCG
TTGCCCTGATGGCTGATTTATTTTTTCGGCATCGACACGTAGC
```

OH⁻
5min. RT

```
AACGGGACTACCGACTAAATAAAAAAGCCGTAGCTGTGCATCG
```

2.Annealing

Primer
TGCCCTGATG

3min. 100 °C → 0 °C

```
AACGGGACTACCGACTAAATAAAAAAGCCGTAGCTGTGCATCG
TGCCCTGATG
```

3.Labelling

dG
dT *low conc. deoxynucleotides*
dC

dA *labelled deoxynucleotide*
m. T7 pol .(enzyme)

5min. RT

```
AACGGGACTACCGACTAAATAAAAAAGCCGTAGCTGTGCATCG
TGCCCTGATGGCTGATTTAT
```

4.Termination

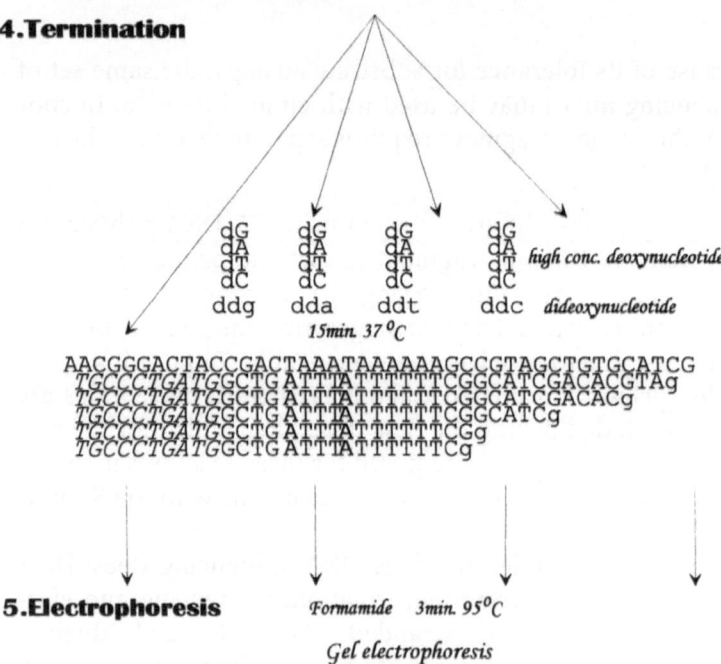

```
dG        dG        dG        dG
dA        dA        dA        dA    high conc. deoxynucleotides
dT        dT        dT        dT
dC        dC        dC        dC
ddg       dda       ddt       ddc   dideoxynucleotide
        15min. 37 °C
```

```
AACGGGACTACCGACTAAATAAAAAAGCCGTAGCTGTGCATCG
TGCCCTGATGGCTGATTTATTTTTTCGGCATCGACACGTAg
TGCCCTGATGGCTGATTTATTTTTTCGGCATCGACACg
TGCCCTGATGGCTGATTTATTTTTTCGGCATCg
TGCCCTGATGGCTGATTTATTTTTTCGg
TGCCCTGATGGCTGATTTATTTTTTCg
```

5.Electrophoresis

Formamide 3min. 95°C

Gel electrophoresis

Fig. 2.1. Principal scheme of double-stranded DNA sequencing using T7 polymerase

Procedures for DNA sequencing described here are based on our current experience in applying the following protocols:

- LKB Macrophor Sequencing System, Laboratory Manual (1985),

- Pharmacia Biotech T7 Sequencing Kit Instructions (1993).

2.1
Preparation of Template DNA

Materials

- Centrifuge (refrigerated, operating up to 3000 g)   Equipment
- Microfuge (Beckman)
- Water bath

- minimal-medium plates   Media M9
  10×M9 salt solution:

Dissolve 60 g of Na_2HPO_4, 30 g of KH_2PO_4, 10 g of NH_4Cl and 5 g of NaCl in deionized water, fill up to 1 l and autoclave.

For 0.5 l of M9 medium, sterilize the following components separately, cool down to approx. 50 °C and mix under sterile conditions:
450 ml of autoclaved deionized water with addition of
7.5 g of Bacto agar (Difco 0140–01–0)
50 ml of autoclaved 10xM9 salt solution
0.5 ml of autoclaved $MgSO_4$ solution, (1 mol/l)
0.05 ml of autoclaved $CaCl_2$ solution, (1 mol/l)
2.5 ml of sterile filtered vitamin B1 solution, (1 mg/ml)
5 ml of autoclaved glucose solution, 20 % (w/v)
Pour 25 ml of each solution into sterile Petri dishes and allow to solidify.
- 2-YT liquid medium
Dissolve 15 g of Bacto tryptone (Difco 0123–01–1), 10 g of Bacto yeast extract (Difco 0127–01–7) and 5 g of NaCl in deionized water, make up volume to 1 l and autoclave.
- H medium plates and H top agar
Dissolve 10 g of Bacto tryptone (Difco) and 8 g of NaCl in deionized water, make up volume to 1 l and autoclave. For pouring plates, add 12 g of Bacto agar (for top agar: 8 g of Bacto agar) and autoclave. Pour 25 ml of each component into sterile Petri dishes and allow to solidify.

- 2-YT solid medium
  2-YT liquid media add
  1.5 % Bacto agar (Difco)
  100 mg/l ampicillin.

Buffers
- $CaCl_2$ solution
  50 mM $CaCl_2$, autoclaved
- Glycerol/$CaCl_2$ solution
  15 % (v/v) glycerol
  50 mM $CaCl_2$, autoclaved
- Tris-HCl solution
  50 mM Tris-HCl, pH 7.2, autoclaved
- IPTG solution
  100 mM isopropyl-β-D-thio-galactoside sterilized by filtration
- X-gal solution
  2 % (w/v) 5-bromo-4-chloro-3-indolyl-β-D-galactoside in
  dimethylformamide
- PEG/NaCl solution
  20 % PEG 6000
  2.5 M NaCl
- TE buffer
  10 mM Tris-HCl
  1 mM EDTA, pH 8.0
- Phenol
  phenol saturated with TE buffer
- Solution 1
  50 mM glucose
  25 mM Tris-HCl pH 8.0
  10 mM EDTA
- Solution 2
  1 % SDS
  0.2 M NaOH
- Solution 3
  3 M Na acetate, pH 5.0
- Solution 4
  4 M Na acetate, pH 8.0

Procedure

Preparation of Competent *E. coli* Cells

1. Remove a single *E. coli* JM103 colony from an M9 minimal medium plate and inoculate with it 5 ml of 2-YT liquid medium. Incubate the culture overnight at 37 °C with shaking.

2. Inoculate 40 ml of 2-YT liquid medium with 0.4 ml of a fresh overnight culture and incubate for approx. 2 h at 37 °C with shaking until $A_{550} = 0.2$.

3. Centrifuge the cells for 10 min at 3000 g in a precooled rotor.

4. Resuspend the pellet in 20 ml of an ice-cold $CaCl_2$ solution and let it stand on ice for 30 min.

5. Centrifuge the cells as described above (step 3).

6. Resuspend the pellet in 4 ml of an ice-cold sterile $CaCl_2$ solution. These competent cells can be stored on ice for up to 24 h. The cell sediment can also be resuspended in an ice-cold glycerol/$CaCl_2$ solution and stored in 300 µl aliquots at -70 °C. Transfection efficiency may decrease during this procedure.

Transfection

1. Inoculate 20 ml of 2-YT liquid medium with 1 drop of *E. coli* JM103 from a fresh overnight culture and shake at 37 °C.

2. Mix the ligated DNA with a Tris-HCl solution to give a final volume of 50 µl (e.g., 5 µl of ligation mixture and 45 µl of Tris-HCl solution).

3. Mix the DNA in a sterile test tube on ice with 300 µl of competent *E. coli* cells and incubate on ice for 40 min.

4. Warm the test tubes in a water bath to 42 °C for 3 min, then put on ice.

5. Melt H top agar, cool down to 42 °C.
   For each transfection experiment, mix:
   200 µl of exponentially growing *E coli* cells (culture from 1., A_{550} approx. 0.3), 40 µl of IPTG solution, 40 µl of X-gal solution. Wear gloves!

6. To each test tube containing competent cells, add under sterile conditions:
   - 270 µl of JM103/IPTG/X-gal mixture,
   - 3 ml of H top agar.
   Mix carefully by rotating and pour immediately onto a warmed H medium plate.

7. After solidification of top agar, invert plates and incubate at 37 °C.

Preparation of Single-Stranded Template DNA

1. Pick a single recombinant plaque, transfer it into 2 ml of 2-YT medium using a sterile glass capillary, and shake at 37 °C overnight.

2. Centrifuge at 5000 g for 30 min at 4 °C.

3. Transfer supernatant to a clean 1.5 ml microcentrifuge tube. Take care not to carry over any cells, leave some of the supernatant behind if necessary. Discard pellet.

4. Add 0.2 volumes of PEG/NaCl solution to the supernatant. Mix well, then let stand for 1 h at 4 °C.

5. Centrifuge at 5000 g for 20 min. Discard supernatant.

6. Centrifuge at 5000 g for 5 min and remove all the remaining PEG/NaCl with a glass capillary.

7. Resuspend the viral pellet in 500 µl of TE buffer.

8. Centrifuge for 5 min in a microcentrifuge to remove the remaining cells. Transfer supernatant to a fresh microcentrifuge tube.

9. Add 200 µl of PEG/NaCl to the microcentrifuge tube. Mix well, then let stand for 15 min at ambient temperature. Alternatively, leave overnight at 4 °C.

10. Centrifuge for 5 min, discard supernatant.

11. Centrifuge for 2 min. Carefully remove all remaining traces of PEG with a glass capillary. Wipe off with a tissue any traces of PEG on the mouth of the tube.

12. Resuspend the viral pellet in 500 µl of TE buffer.

13. Add 200 µl of phenol. Vortex for 15–20 s.

14. Allow the tube to stand for 15 min at ambient temperature. Vortex for 15 s.

15. Centrifuge for 3 min.

16. Transfer the upper (aqueous) layer to a fresh microcentrifuge tube.

17. Repeat steps 13–16.

18. Add 500 µl of diethyl ether, mix well and discard the top layer.

19. Repeat step 18 three times.

20. Extract the aqueous phase with 500 µl of chloroform (twice). Discard the lower organic layer.

21. Add 40 µl of 3 M NaCl and 1 ml of ethanol to each tube with stirring.

22. Spin the tubes in a microcentrifuge for 15 min.

23. Wash each pellet with 1 ml of cold ethanol. Pour off ethanol and allow to drain until dry.

24. Redissolve each pellet in 10 µl of TE buffer.

Transformation

1. Prepare competent *E. coli* cells as described in Chapter 1.

2. Mix DNA in a sterile test tube on ice with 100 µl of competent *E. coli* cells and incubate on ice for 20 min.

3. Warm the test tubes in a water bath to 42 °C for 1 min, then put on ice.

4. Add 900 µl of warm 2-YT media and shake for 1 h at 37 °C.

5. Spread an appropriate quantity of cells (about 200 µl per one plate) onto selective media (solid 2-YT). Incubate overnight at 37 °C.

Preparation of Double-Stranded Plasmid DNA

1. Using a loop, transfer the bacteria to a sterile tube containing 3 ml of 2-YT with ampicillin 100 mg/l.

2. Incubate the culture overnight at 37 °C with shaking.

3. Centrifuge twice at 8000 g for 10 min at 4 °C in 1.5 ml microcentrifuge tube.

4. Resuspend the cell pellet in 200 µl of solution 1.

5. Add 200 µl solution 2. Vortex. Allow the tube to stand for 5 min at ambient temperature.

6. Add 150 µl of solution 3. Mix well, then let stand for 15 min at −20 °C.

7. Warm 5 min at 65 °C.

8. Add 150 µl of solution 4. Mix well, then let stand for 15 min at −20 °C.

9. Centrifuge at 10 000 g for 30 min.

10. Transfer supernatant to a clean 1.5 ml microcentrifuge tube.

11. Add 500 µl of isopropanol. Mix well.

12. Spin the tubes in a microcentrifuge for 15 min.

13. Wash each pellet with 1 ml of cold 70 % ethanol. Pour off the ethanol and let the pellets drain until dry.

14. Redissolve each pellet in 100 µl of TE buffer.

15. Add 2 µl of RNase (10 mg/ml).

16. Incubate for 1 h at 37 °C.

17. Extract aqueous phase with 100 µl of chloroform.

18. Centrifuge for 3 min.

19. Transfer the upper (aqueous) layer to a fresh microcentrifuge tube.

20. Add 5 µl of 5 M NaCl and 250 µl of ethanol to each tube with stirring.

21. Spin the tubes in a microcentrifuge for 15 min.

22. Wash each pellet with 1 ml of cold 70 % ethanol. Pour off ethanol and let the pellets drain until dry.

23. Redissolve each pellet in 50 µl of TE buffer.

2.2
DNA Sequencing

Materials

- Macrophor Sequencing System (Pharmacia)   **Equipment**
- Thermostat (150 °C)
- X-ray film (Amersham, Hyperfilm-MP)

- 5× Annealing Buffer   **Buffers**
  200 mM Tris-HCl pH 7.5
  100 mM MgCl
  250 mM NaCl
- Sequencing primer 0.2 A_{260}/ml
- 5× Labeling mix (A)
  7.5 µM dGTP
  7.5 µM dCTP
  7.5 µM dTTP
- Enzyme Dilution Buffer
  20 mM Tris-HCl pH 7.5
  5 mM DTT
  100 µg/ml BSA
  5 % glycerol
- Stop buffer
  0.25 % bromophenol blue
  0.25 % xylene cyanol in
  100 % formamide
- Bind-Silane solution
  10 ml ethanol
  2.3 ml distilled water
  0.2 ml acetic acid
  40 µl Bind-Silane (Pharmacia 1850–251)
- Repel-Silane (Pharmacia 1850–252)
  2 % dimethyldichlorsilane solution in 1,1,1-trichloroethane
- 10×TBE
  1 M Tris
  0.83 M boric acid

10 mM EDTA, pH 8.3
- 6 % acrylamide solution
  75 ml 40 % acrylamide +2 % bis-acrylamide stock solution
  210 g urea
  50 ml 10'TBE
  make up to 500 ml with distilled water
- Gel washing solution
  10 % acetic acid
  10 % ethanol
- Preparation of dNTP/ddNTP termination mixture

Nucleotide solutions	G mixture (μl)	A mixture (μl)	T mixture (μl)	C mixture (μl)
0.5 mmol/l dATP	80	80	80	80
0.5 mmol/l dGTP	80	80	80	80
0.5 mmol/l dTTP	80	80	80	80
0.5 mmol/l dCTP	80	80	80	80
2 mmol/l ddGTP	2	–	–	–
2 mmol/l ddATP	–	2	–	–
2 mmol/l ddTTP	–	–	2	–
2 mmol/l ddCTP	–	–	–	2
5 M NaCl	5	5	5	5
Redist. water	173	173	173	173
Final volume	500	500	500	500

Procedure

Template DNA Denaturation

Note. This first step is necessary only when double-stranded templates are used.

1. Transfer 12 μl of DNA (2–3 μg) into 1.5 ml microcentrifuge tube.

2. Add 8 μl of 0.5 M NaOH. Let stand for 5 min at room temperature.

3. Add 8 μl of 5 M ammonium acetate. Mix well.

4. Precipitate DNA with 100 μl of ethanol.

5. Spin the tubes in a microcentrifuge for 15 min.

6. Wash each pellet with 0.5 ml of cold 70 % ethanol. Pour off ethanol and let the pellets drain until dry.

7. Redissolve each pellet in 6 μl of water.

Annealing Template and Primer

Add to a sterile 1.5 ml microfuge tube:
- 6 μl (2–3 μg) of template DNA,
- 2 μl (0.8 pmol) of M13/pUC sequencing primer,
- 2 μl of 5× Annealing buffer.

Incubate the mixture for 10 min at 65 °C and allow it to cool slowly to room temperature (for single-stranded template DNA).

Incubate the mixture for 3 min at 100 °C, then quickly cool it on ice (for double-stranded template DNA).

Labeling Reaction

1. Add to the annealing mixture:
   - 1 μl 0.1 M DTT,
   - 2 μl diluted labeling mix,
   - 1 μl labeled dATP; 10 μCi with specific activity 300–3000 Ci/mmol; $[\alpha^{32}P]$dATP, $[\alpha^{33}P]$dATP or $[\alpha^{35}S]$dATPαS (Amersham) labeled nucleotides can be used;
   - 2 μl T7 DNA polymerase (Pharmacia).

   Relative properties of ^{35}S, ^{33}P and ^{32}P isotopes (β emitters)

Isotope	^{32}P	^{33}P	^{35}S
Maximum emission energy (MeV)	1.71	0.249	0.167
Half-life (days)	14.3	25.4	87.4

Caution. Use protective shield when handling ^{32}P.

Note. For maximum stability of enzyme, use the stock of T7 DNA polymerase from storage at −20 °C only momentarily to remove an aliquot for dilution. The diluted stock may be kept at 4 °C for up to 1 week and still give good sequencing results.

Enzyme (5 U/μl) dilution table

Number of templates	Volume of T7 Pol.	Volume of dilution buffer	Total volume
2	1.0	4.0	5.0
3	1.5	6.0	7.5

4	2.0	8.0	10.0
5	2.5	10.0	12.5
6	3.0	12.0	15.0
7	3.5	14.0	17.5
8	4.0	16.0	20.0
9	4.5	18.0	22.5
10	5.0	20.0	25.0
11	5.5	22.0	27.5
12	6.0	24.0	30.0

Mix by repeated pipetting (avoiding bubbles) and incubate for 2–5 min at room temperature.

Place tube on ice.

Note. Reading sequences close to the primer.

The general conditions described in this manual should be followed for sequencing from the primer up to 300–400 nucleotides. If one is interested only in the sequences close to the primer (<200 nucleotides), it is possible to dilute the labeling mixture further (make a 1:20 dilution of the stock reagent) and keep both reaction times to 3–5 min. The gel should be run only until the first blue dye has migrated about 80 % of the length of the gel. When reading sequences within 20 nucleotides of the 3' end of primer, it is essential that a sufficient amount of the template DNA and primer be present. It is good practice to double the recommended amount of each for optimal results.

Note. Reading long sequences.

For reading beyond 400 nucleotides, the concentration of the dNTPs in the labeling reaction should be increased three- to fivefold (i.e. use the undiluted labeling reaction) and the labeling reaction lengthened to 5 min.

To prolong reading length you also can decrease concentration of dideoxynucleotides fivefold.

Termination Reactions

1. Have on hand 4 tubes labeled G, A, T and C.

2. Place 2.5 µl of the G termination mix in the tube labeled G. With a fresh tip for each, fill tubes labeled A, T and C with 2.5 µl of the A, T and C termination mixture, respectively. This is done best before beginning the labeling reaction.

3. When the labeling reaction is completed, remove 3.5 µl of it and transfer it to the tube labeled G. Mix thoroughly. Similarly, transfer 3.5 µl of the labeling reaction to the A, T and C tubes. Use a fresh tip for each transfer.

4. Incubate for 15 min in a bath at 37 °C.

5. Terminate the reactions by adding 4 µl of Stop buffer.

6. Immediately before loading the samples onto the sequencing gel, denature the samples by incubating for 3 min at 95 °C in a water bath, then quickly cool on ice and centrifuge briefly. Immediately load 1–2 µl of each sample onto the appropriate slot of a pre-electrophoresed and preheated (55 °C) sequencing gel.

2.3
Gel Electrophoresis

Procedure

Bind-Silane Treatment of Notched Glass Plate

Place a clean notched glass plate, with the small bevelled area facing upwards, in a fume hood. Next using lint-free tissue, spread the Bind-Silane solution evenly over the top surface of the plate. Let the plate dry, and then polish it with a fresh lint-free tissue. Rinse the plate with ethanol and, while it is still wet, polish it with another tissue. Let the plate dry once more before giving it a final polish with another lint-free tissue. If the plate is not going to be used immediately, it should be covered with lint-free paper until used.

Caution. Never use Bind-Silane on a thermostatic plate. The gel will bind to both the thermostatic plate and the notched plate and it will be impossible to remove it for evaluation.

Repel-Silane Treatment of Thermostatic Plate

Place a clean thermostatic plate on a Macromould gel casting table so that the protruding edge of the plastic frame is turned downwards and the water outlet/inlet is at the end farthest from the moveable plate support. Using a lint-free tissue, spread approximately 5 ml of Repel-Silane carefully over the entire top surface of the plate and let it dry for a few minutes. Polish the plate with a fresh lint-free tissue. After polishing, rinse plate with ethanol and, while it is still wet, polish the plate once more. Let the plate dry for a few minutes, then polish with another lint-free tissue. If the plate is not going to be used immediately, it should be covered with lint-free paper until used.

Caution. Before pouring the gel, the Silane-treated plates should not be left in contact with each other for more than 1 min, even if the plates are separated by spacers. If this is allowed to occur, the Bind-Silane will diffuse onto the thermostatic plate and the gel will bind to both plates.

Both Repel-Silane and Bind-Silane are toxic and highly volatile and should be used only in a fume hood.

Gel Electrophoresis and Autoradiography

Caution. Acrylamide is a potent cumulative neurotoxin. To avoid contact with skin and inhalation of dust and vapours, we strongly recommend that all procedures involving unpolymerized acrylamide, powder or solution, are performed under a fume hood and that disposable gloves are worn.

1. We recommend using 6% polyacrylamide gels with a 0.2–0.4 mm thickness gradient with a Macrophor Sequencing System.
   Immediately before pouring the sequencing gel to 50 ml of 6% acrylamide solution add:
   500 µl of 10% ammonium persulphate solution,
   50 µl of TEMED.

2. The recommended gel electrophoresis conditions are 2 kV, approx. 20–40 mA, 55 °C.
   Dye Migration in Polyacrylamide Denaturing Gels

Gel (%)	Bromophenol Blue (Size of the fragments)	Xylene Cyanol (nucleotides)
5.0	35	140
6.0	26	106
8.0	19	75
10.0	12	55
20.0	8	28

3. After electrophoresis, place the sequencing gel into gel-washing solution for 20 min in order to remove urea.

4. Dry for approx. 1 h at 150 °C.

5. Lay an X-ray film on the dried gel and expose overnight (longer or shorter exposure may be necessary).

Troubleshooting

• Specific bands are very light in intensity or missing, or film is blank
- DNA preparation may be bad, try the control DNA.
- Primer does not anneal to template, primer site eliminated during cloning step.
- Labeled nucleotide too old.
- Some components missing.
- Enzyme lost activity.

• Gel sticks to the thermostatic plate
- Wash thermostatic plate and notched glass plate carefully immediately before silane treatment.
- Decrease used amount of Bind-Silane solution.
- Do not keep thermostatic plate and notched glass plate together for long time.

• Bands smeared
- Contaminated DNA preparation, try control DNA.
- Gel may be bad. Gels should be cast with freshly made acrylamide solution and should polymerize within 15 min of pouring.
- Gel run too cold. Sequencing gels should be run at 50–55 °C.
- Gel dried too hot or not flat enough to be evenly exposed to film.
- Samples are not denatured. Make sure samples are always heated immediately prior to loading on gel.

• Sequence faint near the primer
- Insufficient DNA in the sequencing reaction. A minimum of 1 μg of single-stranded DNA and 3–4 μg plasmid is required for sequencing close to primer. Try increasing the amount of DNA.
- Insufficient primer. Use a minimum 0.5 pmol. Primer to template mole ratio should be 1:1 to 1:5.

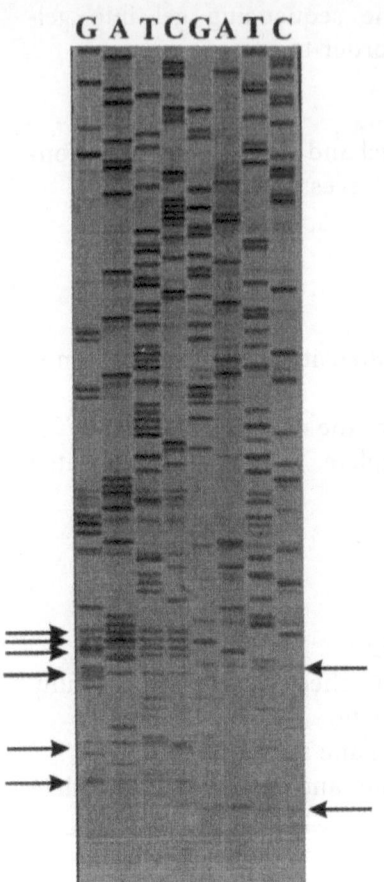

G A T C G A T C

Fig. 2.2. Parallel bands are present in all four lanes close to primer and they are shifted by 'half-step'. Labeling reaction artifacts. *Arrows* indicate nonspecific bands

- A background of even but high intensity is obtained, such that specific bands become hard to distinguish
- Contaminated DNA preparation, try control DNA.

- One or more of the tracks is too faint or too short
- This is due to incorrect dNTP/ddNTP ratios; tracks that are too faint are usually caused by not having enough dideoxy-nucleoside triphosphate present, while short tracks are caused by having too much present.

- Parallel bands are present in all four lanes close to primer and they are shifted by a 'half-step' (Fig. 2.2)
- Labeling reaction artifacts. The template/nucleotide ratio was incorrect, resulting in the synthesis of longer labeled fragments. This reduces the relative number of smaller chain-

Fig. 2.3. Parallel bands are present in all four lanes above 200 bases from the primer. The enzyme may have lost activity. *Arrows* indicate nonspecific bands

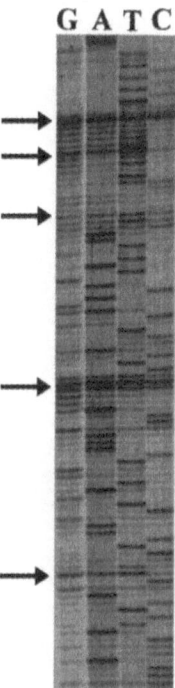

terminated fragments, producing weaker band intensities in the lower section of the gel. Increase the amount of template and primer by a factor of two. Alternatively, decrease the nucleotide concentration in the labeling reaction. The labeling step should not be run warmer than 20 °C or longer than 5 min.

- Parallel bands are present in all four lanes above 200 bases from the primer
- The enzyme may have lost activity (Fig. 2.3). Increase the amount of enzyme in the reaction twofold.

Caution. High concentration of glycerol in the electrophoresis sample can distort sequence pattern in the top of gel.

- The template may have strong secondary structures that cause the polymerase to pause. Following the annealing step, incubate the primer/template solution at 60–70 °C for 4 min, return the solution to room temperature for 2 min and then immediately proceed to the labeling reaction. When sequencing double-stranded templates, it may help to include 20 % dimethyl sulfoxide in the annealing reaction.

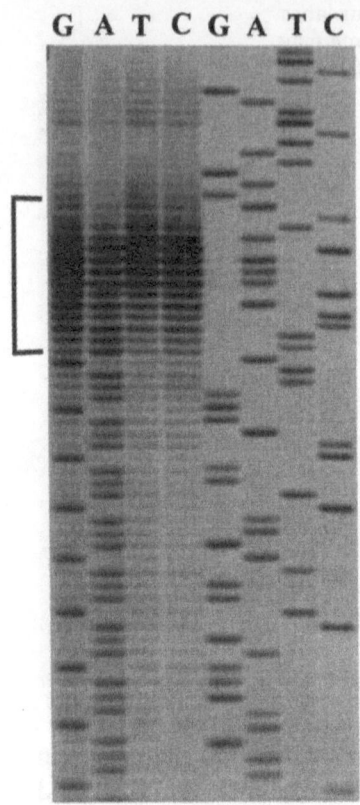

G A T C G A T C

Fig. 2.4. Parallel bands are present in all four lanes above 200 bases from the primer. The template has a GAAA sequence repeated 22 times . *Square bracket* indicates nonspecific bands

- The template may have long repeated sequences that cause the polymerase to pause (Fig. 2.4). Increase distance between primer annealing site and repeated sequence. Try 'READ LONG' conditions.

● Some bands faint (Fig. 2.5)
- Termination reaction time too long. If the termination reaction is allowed to continue too long, the synthesized DNA may be degraded at specific sequences by extraneous nuclease. Try reducing the termination time.

● Band compression masks the correct sequence in a particular region of the gel (Fig. 2.6)
- Fragments differing in size by one or a few nucleotides migrated with similar mobilities, because residues had formed stable intrastrand secondary structures which were not fully denatured during electrophoresis. Repeat the sequencing reac-

Fig. 2.5. Some bands faint. Termination reaction time too long. *Square bracket* indicates unreadable area

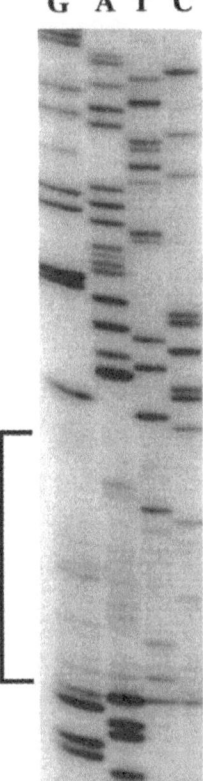

tions with analogs, 7-deaza dGTP and 7-deaza dATP, less able to form intrastrand base pairs.

● Each band in the pattern appears as a doublet or triplet
– This occurs if the primer has a heterogeneous 5' end.

G A T C

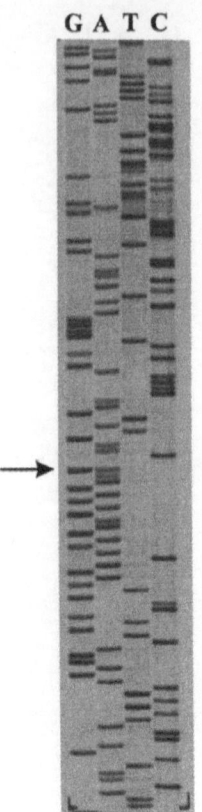

Fig. 2.6. Band compression masks the correct sequence in a particular region of the gel. Fragments differing in size by one or a few nucleotides migrated with similar mobilities. *Arrow* indicates shifted band

References

Birnboim H, Doly J (1979) A rapid alkaline extraction procedure for screening recombinant plasmid DNA. Nucl Acids Res.: 7, 1513–1523

Davies RW (1982) DNA Sequencing. In: Rickwood D, Hames B D (eds) Gel electrophoresis of nucleic acids: A practical approach. IRL Press, Oxford, pp 117–172

Maxam AM, Gilbert W (1980) Sequencing end-labeled DNA with base specific chemical cleavages. In: Grossman L, Moldave K (eds) Nucl Acids. Methods Enzymol, vol 65. Academic Press, London, pp 499–599

Messing J, Bankier AT (1989) The use of single stranded DNA phage in DNA sequencing. In: Howe CJ, Ward ES (eds) Nucleic Acids Sequencing: a practical approach. IRL Press, Oxford, pp 1–36

Sambrook J, Fritsch EF, Maniatis T (1989) Molecular cloning: a laboratory manual. Cold Spring Harbor Laboratory

Sanger F, Nicklen S, Coulson AR (1977) DNA sequencing with chain-terminating inhibitors. Proc Natl Acad Sci USA 74, 5463–5467

Chapter 3

RNA In Vitro Synthesis by Phage T7 DNA-Dependent RNA Polymerase

ELITA AVOTA[*][1] AND NORMUNDS LICIS

Introduction

The purpose of this chapter is to introduce beginners in molecular biology to RNA transcription by phage T7 DNA-dependent RNA polymerase. The work outlined here includes the transcription procedure of plasmid vectors or PCR-amplified DNA templates, the purification and identification of RNA products by sequencing with reverse transcriptase.

A powerful approach for the analysis of gene structure and function has become possible with the development of RNA synthesis in vitro from cloned DNA templates. The RNA polymerases encoded by the *Salmonella typhimurium* bacteriophage SP6 and *Escherichia coli* phage T7 have characteristics suitable for such a system. These enzymes are very efficient, exhibit stringent promoter specificity (Ikeda and Washamana 1992) and are able to copy long heterologous DNA sequences (Ling and Risman 1989). However, they are sensitive regarding the structure of the 3'-ends of DNA template: transcription of templates prepared by digestion with restriction enzymes that leave protruding 3' ends resulted in the production of RNA with an extraneous sequence from a noncoding template (Schenborn and Mierendorf 1985). Sharp and Konarskaya (1990) reported also that the T7 RNA polymerase is able to synthesize short RNA molecules from the NTP mixture in the absence of endogenous template (de novo synthesis).

Phage T7 RNA polymerase consists of a single polypeptide chain of 98 856 daltons and recognizes its own promoter DNA

Principles and Applications

[*] Corresponding author: Elita Avota; phone: +371–2–421796; fax: +371–2–427521; e-mail: eliavota@biomed.lu.lv
[1] University of Latvia, Biomedical Research and Study Centre, Ratsupites Str. 1, LV1067 Riga, Latvia

sequence with very high specificity. During infection of *E. coli* cells, the late genes of bacteriophage T7 are transcribed by an RNA polymerase that is specified by one of the early phage genes (Mackie 1988). The speed of RNA chain growth is 230 bases per second which is five times faster than in the case of host *E. coli* RNA polymerases (McAllister and Carter 1980). Transcription of the late genes is regulated and proceeds in two overlapping temporal events: class II genes are transcribed from 6 until 15 min after infection, while class III genes are transcribed from 6–8 min after infection until beginning of lysis (Joho and Gross 1990). Purified T7 RNA polymerase discriminates between the class II and class III promoters in vitro as a function of variables that alter the stability of the DNA helix (Osterman and Coleman 1981). T7 class II and class III promoters have the consensus sequence of 22–23 bp in length and they are similar to the sequences of other bacteriophages, e.g. T3, SP6 (Fig. 3.1).

The sequence of T7 promoter region from +2 to −17 contains all specific information necessary for transcription initiation with T7 RNA polymerase. The region of 8 bp (−6 to +2) is the site where polymerase binds to the DNA template. The binding occurs on the ss (positive chain) open frame. Nevertheless, the

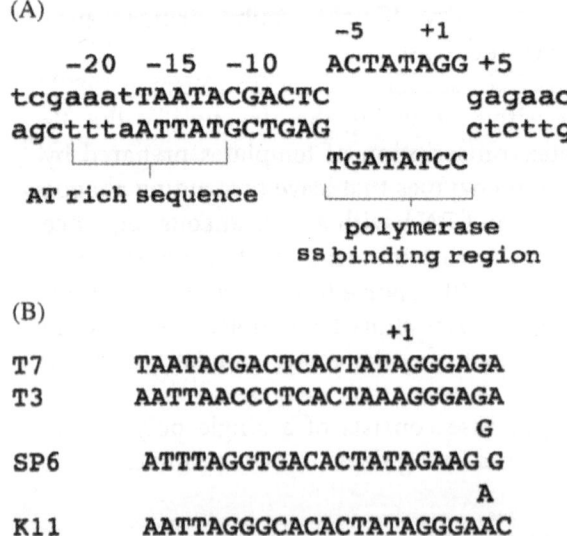

Fig. 3.1. A The sequence of T7 RNA polymerase promoter and its flanking region (Osterman and Coleman 1981). B Comparison of promoter sequences for bacteriophages T7, T3, SP6 and K11. (Klement 1990)

transcription initiation requires the double-stranded form of DNA (Krupp 1988). The AT-rich sequence of the promoter is essential for higher transcription efficiency because it stabilizes the transcription initiation complex. The sequences of different bacteriophage promoters have a high level of homology (Fig. 3.1B). On the other hand, each kind of polymerase is able to initiate the transcription only from the host phage promoter. T7 RNA polymerase has been demonstrated to have the highest specificity in cell-free systems.

Final conclusion: there are two important regions in the sequence of active bacteriophage promoters: polymerase binding site (positions from −6 to +2) and transcription initiation site (positions from +4 to +3).

The use of recombinant plasmids as transcription vectors is a simple and straightforward method for RNA synthesis (Fig. 3.2). Most of inserted genes can be obtained as full-size sense or antisense RNA transcripts and the yields can reach several hundred transcripts per molecule of template DNA. The use of synthetic

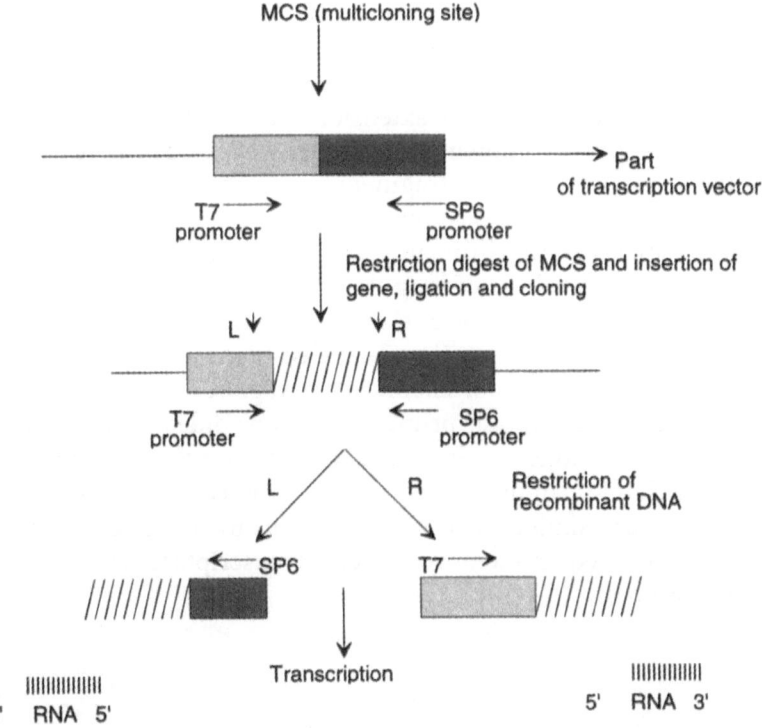

Fig. 3.2. RNA synthesis from transcription vectors

DNA templates for the synthesis of relatively short RNA fragments using T7 polymerase has several advantages. First, the unwanted 5' flanking- sequences that are often introduced during cloning procedure can be eliminated. Second, the sequence at the 3' end of the transcript is not limited by the need for a restriction site to linearize the template. Finally, the time-consuming cloning, sequencing and plasmid preparation can be avoided (Krupp 1988).

The use of transcription products – ssRNAs, as asymmetric probes leads to tenfold increased sensitivities in Southern and Northern blotting. Also, the rather long probes required for genomic sequencing can be prepared. The RNAs can be used as templates for translation in vitro and in vivo. The expression of specific genes can be controlled by the injection of anti-sense RNA into the cell. Transcription in vitro includes substrate synthesis for studies of RNA splicing, tRNA processing and self splicing RNAs, analysis of catalytic RNAs.

Sequencing of Transcription Products

The basic techniques of RNA sequencing are: two-dimensional separation methods (RNA fingerprint), gel sequencing of 5' or 3' end-labeled RNAs or primer-extension sequencing.

A number of two-dimensional systems have been described for the separation of RNA fragment mixtures labeled either in vitro or in vivo. The most common systems use cellulose and gel supports for HVE and chromatography (Howe and Ward 1989).

For sequence analysis of long RNA molecules methods based on electrophoresis in denaturing polyacrylamide gels are generally used. Here, the 3' or 5' end-labeled RNAs are subjected to base-specific enzymatic or chemical cleavage reactions. The use of RNA as template in primer-extension dideoxynucleotide sequencing reactions was introduced soon after the widespread use of of chain-termination sequencing protocols for DNA. The techniques are similar to the reactions using DNA templates, except for the requirements for reverse transcriptase (Howe and Ward 1989).

Materials

– Electroelution chamber (Pharmacia, 2014–001)   **Equipment**

– AMV reverse transcriptase (10 U/μl, Pharmacia, 27–0922–02)   **Reagents**
– T7 RNA Polymerase (20 U/μl, Fermentas, EPO112)
– Ribonuclease inhibitor (35 U/μl, Fermentas, EOO311)
– T4 polynucleotide kinase, (15 U/μl Fermentas, EK0032)

10× Transcription buffer   **Buffers**
400 mM Tris-HCl, pH 7.5
60 mM MgCl$_2$
40 mM spermidine
– 5× SB-Mg (buffer for synthesis without Mg)
  60 mM NH$_4$Cl
  10 mM Tris-acetate, pH 7.4
  6 mM β-mercaptoethanol
– NTP solutions for RNA sequencing
  3.75 mM each of dNTP
  1 mM ddCTP
  2 mM ddATP
  3 mM ddTTP
  1.5 mM ddGTP
– Stop mix
  94 % formamide
  36 mM Tris-borate, pH 8.0
  36 mM boric acid
  0.8 mM EDTA

Procedure

DNA Template Preparation

Carry out the preparation of the vector plasmid by standard methods so that DNA is free of RNA and protein contamination (Maniatis and Fritsch 1982; see also Chap. 2 by Jankevics).

The plasmid containing T7 polymerase promoter is completely linearized by the appropriate restriction endonuclease to obtain RNA of a specified length. Certain precautions must be taken when using the restriction endonucleases:

- Enzymes that result in DNA 3'-overhang must be avoided to eliminate synthesis of longer transcripts; only blunt and 5'-protruding ends of template are appropriate for transcription; therefore, we recommend that plasmids should not be linearized with the following enzymes listed here: AatII, ApaI, BanII, BanII, BglI, Bsp1286, BstXI, CfoI, HaeII, HgiAI, HhaI, KpnI, PstI, PvuI, SacI, SacII, SfiI, SphI. If it is not possible to escape the use of the above restriction enzymes, then blunt ends can be prepared using Klenow fragment DNA polymerase or T4 DNA polymerase for filling.
- A complete restriction of the DNA template is obligatory for an effective transcription of a desired sequence because T7 RNA polymerase prefers the unlinearized plasmid as a template and this results in long transcripts which dominate in the transcription pattern.
- Linearized DNA templates are preferred to templates which are cut in to numerous fragments; the T7 RNA polymerase, to some extent, recognizes the free ends of DNA fragments as initiation sites for transcription: this results in a mixture of RNA transcripts of various sizes and sequences.

The procedure after DNA restriction is as follows:

1. Extraction of the restriction mixture with an equal volume of phenol:chloroform:isoamyl alcohol (24:24:1).

2. Threefold extraction of upper phase with 1 vol of chloroform.

3. Precipitation of DNA from the solution in 0.3 M sodium acetate, pH 8.0, with 3 vol of ethanol at −70 °C for 30 min (very effective is the precipitation with 1/10 volume of 100 mM spermidine at 0 °C for 30 min in the presence of 0.3 M sodium acetate, pH 8.0, and 12 mM $MgCl_2$).

4. Centrifugation for 20 min at 10 000 rpm.

5. Washing the pellet twice with 70 % ethanol and drying under vacuum.

6. Dissolving DNA in double-distilled deionized water (final concentration – 1 µg/µl).

Synthesis of Large Amounts of RNA

1. Composition of transcription mixture (final volume = 100 µl) **Standard Protocols**
   10 µl 10× transcription buffer
   10 µl 100 mM DTT (freshly prepared)
   100 U ribonuclease inhibitor
   25 µl 8 mM each of ATP, GTP, CTP, UTP (final concentration of 2 mM)
   2 µl (2 µg) linearized template DNA or PCR product, containing T7 promoter and DNA sequence of needed length)
   add H_2O to final volume 100 µl
   100 U T7 RNA polymerase

Caution. The order of the transcription mixture preparation must be strictly considered to avoid DNA template precipitation at high concentration of spermidine in 10× transcription buffer at the beginning of mix preparation.

2. Incubate mixture at 37 °C for 1 h. Yields of 4–6 µg RNA can be obtained using the conditions described above.

3. Extraction of RNA transcript from the reaction mixture;
   - remove DNA template with RNase-free Dnase I (0.2 mg/ml or 1 U of enzyme per 1 µg DNA template) at 37 °C for 10 min;
   - extract the transcription mixture with equal volume of phenol saturated with water (acidic phenol);

Note. After this step, transcription mixture can be loaded on PAAG, followed by electrophoresis to separate the transcripts and then isolate the target RNA band from gel slices (see below).

   - precipitation of RNA from an aqueous phase containing 0.3 M Na acetate, pH 4.9 with 3 vol of ethanol; centrifugation at 10 000 rpm for 20 min,
   - wash the pellet twice with 70 % ethanol;
   - dry pellet and dissolve in 20 µl of double-distilled water to a final volume corresponding to 1/4 to 1/5 of initial volume of transcription mixture.

Purification of RNA by PAGE Electrophoresis

This example of nucleic acid elution procedure is adapted from the manual of the Pharmacia electroelution chamber Extraphor.

RNA is mixed with 0.2 vol 0.01 % each of xylene cyanol and bromophenol blue dyes in deionized formamide, heated at 95 °C for 3 min and cooled rapidly at −70 °C, and then loaded on 8 % PAAG containing 7 M urea. After electrophoresis the RNA is visualized on the wet gel by staining with ethidium bromide. The band is then excised from gel and RNA electroeluted from the slices into 4 M NH$_4$Cl, pH 5.5, using Extraphor electrophoretic concentrator at 100 V, 20 mA for about 1 h. To use Extraphor, fill the buffer chambers with running buffer, then place the gel slices in the central core and cover with buffer. Inject 50 µl of salt solution of high molarity (4 M NH$_4$Cl) underneath the buffer in the V-shaped channel. Apply electrical field. When current is applied, RNA migrates out of the slices, enters the V-shaped channel and concentrates inside the highly conductive high-salt region, where the ions in solution provide the major charge transport and the rapid migration of RNA is reduced. After elution, RNA is recovered from the high-concentration salt solution by ethanol precipitation and dissolved in water. RNA from the 100 µl transcription mixture is isolated from the gel and dissolved in 50 µl distilled water to obtain the final concentration of RNA of 0.1 µg/µl.

Primer-Extension Sequencing of RNA with AMV Reverse Transcriptase

5'-End-Labeling of Oligonucleotide Primer

The 5'-end labeling of the oligonucleotide primer can be performed before the sequencing by the following procedure.

Ten picomoles of the synthetic oligonucleotide primer is 5'-end labeled with 10 pM of [γ-^{32}P] ATP in a 10 µl reaction using T4 polynucleotide kinase under standard conditions (Maniatis 1982). The reaction mixture is sequentially extracted with phenol-chloroform and ether, and dried in a vacuum. The labeled primer is resuspended in 20 µl of water.

Sequencing Procedure

This method has been developed according to Hartz's (1988) extension inhibition analysis using dideoxynucleotides in the reaction to obtain the sequencing pattern (Stern and Wilson 1986).

Annealing mixture:

2 µl 5× SB-Mg

1 µl (2 pmol of RNA transcript or 0.1 µg of 135 bases long ssRNA)

6 pmol 5'-^{32}P-end-labeled oligonucleotide primer (made up as given above)

0.5 µl formamide

6.5 µl water for adjustment the final volume of 10 µl

incubate at 65 °C for 3 min

then cool slowly to 42 °C

Add 2 µl SB (+10 mM Mg acetate) buffer to the annealing mixture and divide mixture into four equal parts

Extension reaction:

- 3 µl Annealing mixture
  1 µl of dNTP mix
  1 µl of ddNTP mix
  0.5 U AMV reverse transcriptase
  incubate at 42 °C for 15 min
- Termination of reaction:
  Add 5 µl of stop mixture
  incubate at 95 °C for 2 min
- RNA sequencing gel:

Sequencing products are separated by size on denaturing 6 % PAAG containing 8 M Urea (0.2–1 mm gradient gel). The ratio of acrylamide to bisacrylamide is 19:1. Pre-electrophorese the gel at 800–900 V for 20 min. Rinse urea out of wells and load 1 µl of each sample into wells. Electrophorese samples at 1850 V and 55 °C until desired separation is obtained. After electrophoresis, cover gel with plastic wrap and expose it to an X-ray film at −70 °C overnight.

Results

Short transcripts (135–150 nt) from T7 promoter containing plasmids were used for Northern blot hybridization – Qβ phage replicase amplified RNA sequence identification. They can be labeled by radioactivity by introducing the radioactive ribonucleotide into the transcription mixture. The sequencing of ss RNA transcripts was done to prove the correctness of RNA 5' ends, which is crucial for in vitro amplification by Qβ replicase. To reduce the secondary structure effects on AMV reverse transcriptase sequencing, the introduction of formamide into the sequencing mixture significantly improved the sequencing pattern.

References

Hartz D, McPheeters DS, Traut R, Gold L (1988) Extension inhibition analysis of translation initiation complexes. In: Noller HF Jr, Moldave K (eds) Methods Enzymology 164:419–425

Howe, CJ, Ward ES (1989) Nucleic Acid Sequencing. Oxford University Press, Oxford, pp 30–75

Ikeda RA, Warshamana GS (1992) In vivo and in vitro activities of point mutants of the bacteriophage T7 RNA polymerase promoter. Biochemistry 31:9073–9080

Joho KE, Gross LB (1990) Identification of a region of the bacteriophage T3 and T7 RNA polymerases that determines promoter specificity. J Mol Biol 215:31–39

Klement JF, Moorefield MB, Jorgensen E, Brown JE, Risman S, McAllister WT (1990) Discrimination between bacteriophage T3 and T7 promoters by the T3 and T7 RNA polymerases depends primarily upon a three base pair region located 10 to 12 base-pairs upstream from the start site. J Mol Biol 215:21–29

Konarska MM, Sharp PA (1990) Structure of RNAs replicated by the DNA-dependent T7 RNA polymerase. Cell 63:609–618

Krupp G (1988) RNA synthesis: strategies for the use of bacteriophage RNA polymerases. Gene 72:75–89

Ling M, Risman SS (1989) Abortive initiation by bacteriophage T3 and T7 RNA polymerases under conditions of limiting substrate. Nucl Acids Res 17:1605–1616

Mackie GA (1988) Vectors for the synthesis of specific RNAs in vitro. In: Rodriguez RL, Denhardt DT (eds) A survey of molecular cloning vectors and their uses. Butterworths, New York, pp 253–267

Maniatis T, Fritsch EF, Sambroock (1982) Molecular cloning: a laboratory manual. Cold Spring Harbor, New York, pp 1.25–1.52

McAllister WT, Carter AD (1980) Regulation of promoter selection by the bacteriophage T7 RNA polymerase in vitro. Nucleic Acids Res 8: 4821–4837

Schenborn TE, Mierendorf RC (1985) A novel transcription property of SP6 and T7 RNA polymerases: dependence on template structure. Nucl Acids Res 13:6223–6236

Stern RC, Wilson HF (1986) Protein-RNA interactions. J Mol Biol. 192:101–105

Chapter 4

PCR-Based Site-Specific Mutagenesis

Janis Klovins[*1] and Valdis Berzins

Introduction

The alteration of gene structure through the substitution of specific nucleotides by site-specific mutagenesis is an important tool in modern recombinant DNA technology. Nucleotide changes are necessary not only for the analysis of the structural basis of gene and corresponding protein function, but also for the generation of novel gene products. The availability of the polymerase chain reaction (PCR) in the last decade has enabled the modification of DNA for different needs to be made more rapidly and easily than was previously possible. In the course of mutagenesis the relevant sequence changes can be introduced more readily by chemically synthesized oligonucleotide primers than by manipulating DNA fragments with restriction and ligation enzymes.

Principle and Applications

The introduction of some alterations (deletion, insertion and base changes) in terminal parts of the DNA sequence is quite simple; mutagenesis is carried out by PCR via modified primers (Vallete et al. 1988; Sharrocks and Shaw 1992). The fact that these sequences are mismatched to the template DNA in most cases has little effect on the specificity or efficiency of the amplification. Specificity of synthesis is dependent mainly on the sequence of the 3' end of the primer (Nassal and Rieger 1990). DNA strands initiated by these mutant primers are themselves copied and the changed sequence becomes fixed into the growing populations of PCR product fragments. The principle of

[*] Corresponding author: Janis Klovins, phone: +371−2−421796; fax: +371−2−427521; e-mail: klovins@biomed.lu.lv
[1] University of Latvia, Biomedical Research and Study Centre, Ratsupites Str. 1, LV1067 Riga, Latvia

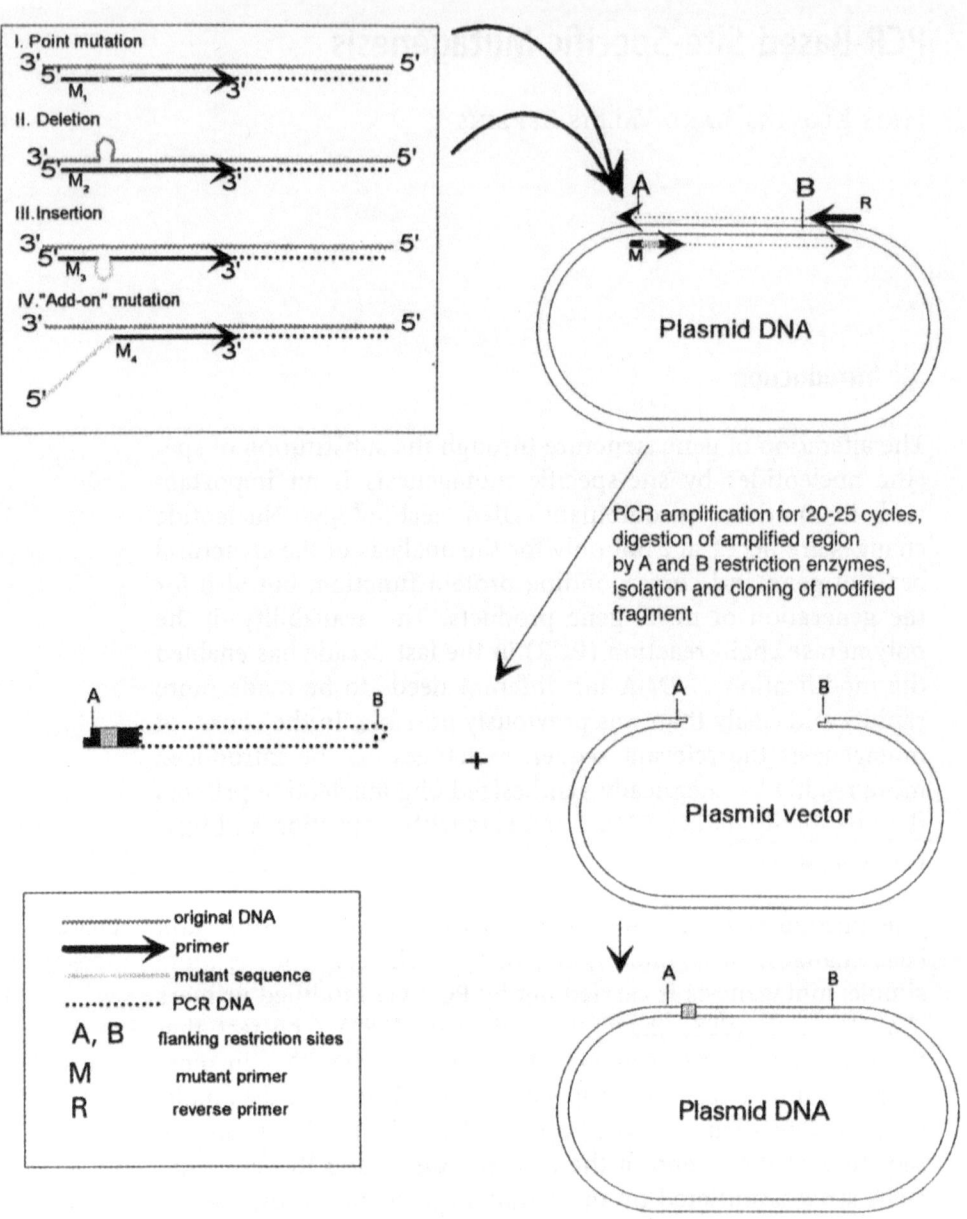

I. Point mutation

II. Deletion

III. Insertion

IV. "Add-on" mutation

A   B
R

Plasmid DNA

PCR amplification for 20-25 cycles,
digestion of amplified region
by A and B restriction enzymes,
isolation and cloning of modified
fragment

+

Plasmid vector

original DNA
primer
mutant sequence
PCR DNA
A, B   flanking restriction sites
M   mutant primer
R   reverse primer

Plasmid DNA

Fig. 4.1. PCR-based introduction of point mutations, deletions, insertions and additional sequences into a specific DNA region which is cloned into a high-copy-number vector. Site for reverse primer (*R*) flanks the one of the restriction site (*B*). The sequence of a mutant primer (*M*) at the 5'-end contains a second restriction site (*A*). The amplified fragments are cut out with the appropriate restriction endonucleases and subcloned into a plasmid vector. The resultant recombinant plasmid contains only the mutagenized DNA region of interest

alteration of DNA is shown in Fig. 4.1; it represents the simplest way for site-directed mutagenesis based on PCR of cloned genes. The products obtained are suitable for cloning into different expression vectors.

However, the use of this method as a general approach for site-directed mutagenesis has limitations because all sequence alterations must be inserted within the primers located at the ends of the target DNA sequence. Therefore the restriction sites must also be located at the ends of the DNA region to permit cloning, and the sites of mutagenesis must be at the terminus near these restriction sites. The insertion of mutations at other sites inside of the amplified gene sequence is therefore not possible.

Using mutagenesis by overlap extension allows the introduction of sequence alterations at any position within the DNA region, not just at the ends of the sequence. This method makes possible the insertion of specific mutations into the nucleotide sequence directly from the cloned gene into its original vector with approx. 100 % efficiency using a few simple steps (Ho et al. 1989). The principle of the method is the generation of two PCR fragments with overlapping ends that can be effectively fused by recombining them in a subsequent PCR reaction. The scheme of this procedure is illustrated in Fig. 4.2A. Specific base or site substitutions, deletion and insertion can also be introduced by this method (Higuchi 1989).

Modifications of this technique have been described. In the method presented by Perrin and Gilliland (1990), only one mutant primer is required and mutations can be introduced into large DNA fragments (up to 1500 bp). In brief (Fig. 4.2B), a long, single-stranded mutant primer is synthesized by PCR by means of a short mutant primer and wild-type sequence as a 5' primer. After removal of unincorporated primers, the large mutant megaprimer is added to a standard PCR mixture that contains a wild-type 3' primer and a catalytic amount of wild-type template. The subsequent PCR generates a mutant DNA fragment of the desired length.

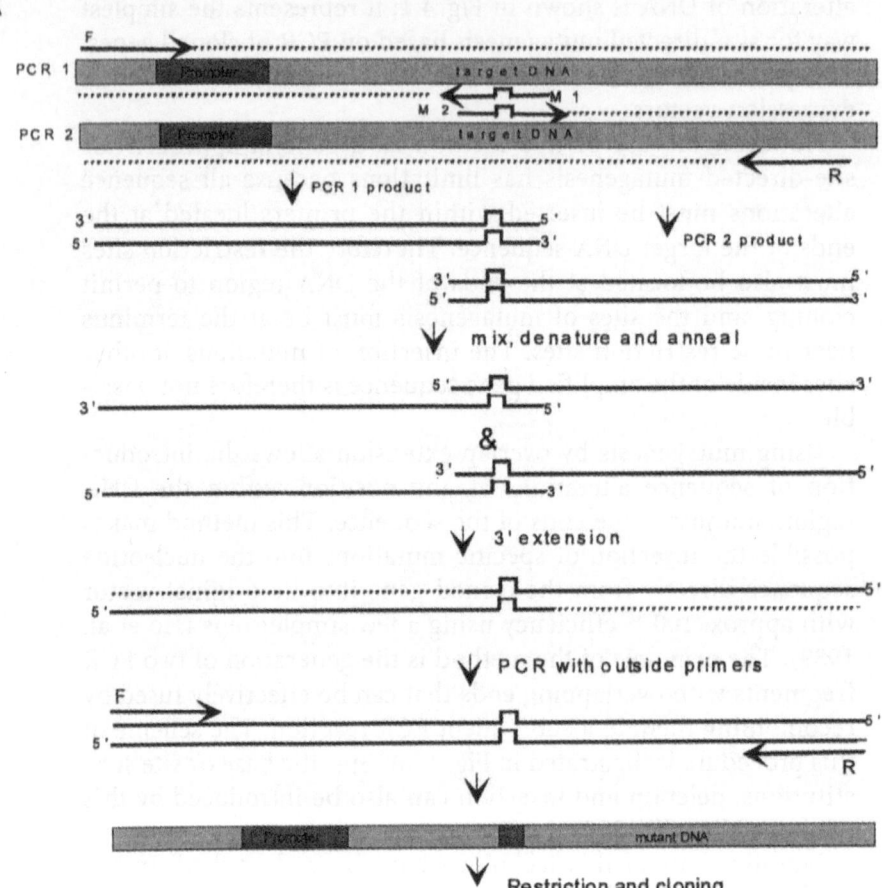

Fig. 4.2A-B. Introduction of mutations in internal part of DNA sequence.
A Two-step site-specific PCR mutagenesis. Sites for two oligonucleotide forward and reverse primers (*F* and *R*) flank the cloning site. In two separate reactions, fragments upstream (*F* and *M1*) and downstream (*R* and *M2*) DNA fragments are PCR-amplified using flanking primers and oligonucleotides containing the mutations (*M1*, *M2*). The PCR products are purified, mixed, denatured and allowed to re-anneal. Some of the molecules recombine through the overlap made by internal primers. DNA chain extension of the recombinants forms the molecules that can be amplified with the flanking primers (*F* and *R*). This product can be digested with the appropriate restriction endonuclease and ligated into a vector. **B** Megaprimer-specific PCR mutagenesis. In first asymmetric PCR reaction single-stranded mutant DNA is generated. This long PCR product is used as mutant primer in the subsequent PCR reaction together with the second standard primer

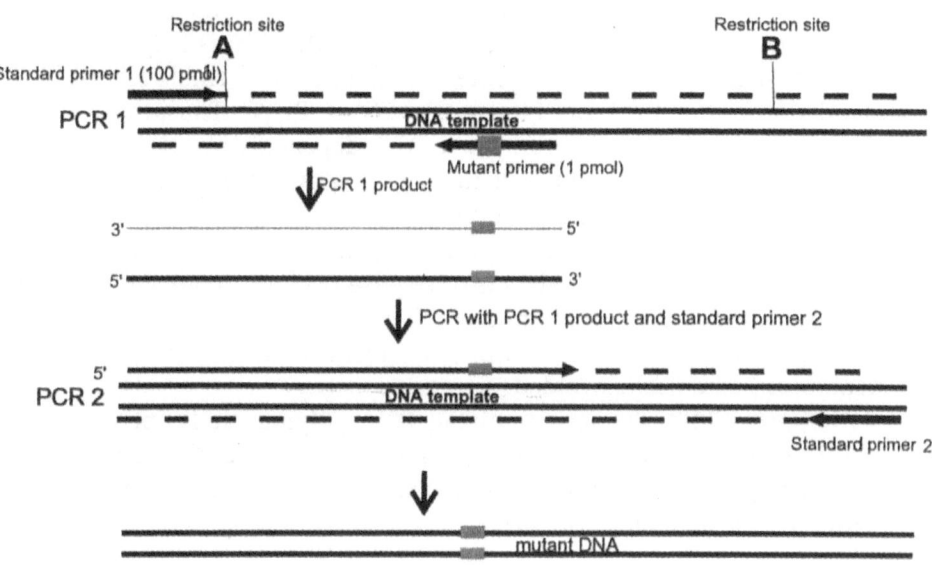

Fig. 4.2B

Primer Design

In general, primers have at least 15 bases of target sequence complementarity that are 3' to any add-on sequences. Single base mismatches can probably occur anywhere except the 3–4 bases from the 3' end. Add-on sequences that provide overlap between PCR products should probably be at least 15 bases long.

To prevent unwanted mutagenesis one should begin with a concentration of template DNA as high as possible and use a minimal number (20–25) of PCR cycles. Since large amounts of initial DNA template are used, gel purification of PCR products is necessary.

Cycle Number and Quantity of Initial DNA Template

Denaturation, Annealing, Elongation Temperature and Time

Normally at 94 °C, the time of denaturation of the template is 60 s; however, to prevent possible DNA damage for short templates denaturation must be shortened to 20 s.

Primer length and sequence are of critical importance in designing the parameters of a successful amplification: the melting temperature (T_m) of a duplex increases both with its length and with increasing of (G+C) content, and a simple formula for calculation of the T_m is $T_m = [4(G + C) + 2(A + T)]$ °C. Thus, the annealing temperature chosen for a PCR depends directly on length and composition of the primers. One should aim at using an annealing temperature (T_a) about 5 °C below the lowest T_m of the pair of primers to be used. 3'-Ends of primers should not be complementary (i.e. base pair), as otherwise primer dimers will be synthesized preferentially to any other PCR product. Primer self-complementarity – the ability to form double-stranded structures – should be avoided.

At approx. 72 °C the activity of Taq DNA polymerase is optimal for elongation, and primer extension occurs at speeds of up to 100 bases/s, therefore, for a short DNA fragments the time of synthesis can be reduced to 30 s.

Unwanted Mutagenesis in PCR

If the mutagenized PCR product is used directly for sequencing or as a template for an in vitro transcription system then low level random mutations which appeared during the amplification should not have a significant effect because most of the molecules have an unaltered nucleotide at any position. However, if the PCR product is later cloned into a vector, than the probability is greater that the cloned molecule will contain a sequence alteration at other than the desired position. As previously described, the Taq DNA polymerase incorporates one incorrect nucleotide for every 9000 nucleotides and causes a frameshift once every 41 000 nucleotides. Misincorporation of nucleotides by Taq DNA polymerase, however, does not always result in sequence alteration in the final product. Since Taq DNA polymerase has no 3' to 5' proofreading exonuclease activity, misincorporated bases can not be removed, however, they can promote termination of the extending DNA chain (Higuchi 1989). If this occurs, the errors will not propagate in subsequent PCR cycles. Sequencing of the cloned PCR product from the same amplification resulted in a different frequency of sequence change: from one unwanted mutation in 400 bases sequenced to one in 3800 bases sequenced (Ho et al. 1989). The frequency of unwanted mutagenesis is dependent on the number of cycles, the conditions of amplification as well as on the sequence being amplified. Therefore, reaction conditions at dNTP concentrations between 50 and 200 µM (higher concentration can promote misincorporation) and at primer annealing temperatures as high as possible (misincorporation at high T_m promotes chain termination) are recommended. To prevent DNA damage that could promote more misincorporation, the total reaction time at high temperatures should be as short as possible and minimal PCR cycle times must be used. The accumulation of mutations in the PCR product is proportional to the number of replications of the DNA, and an optimum number of 20–25 cycles of PCR is most used. To obtain a stable yield, it has been recommended to use an initial template concentration up to 1 µg. Finally, all the cloned constructs made by PCR should be sequenced to verify that no unwanted mutations have occurred.

High fidelity of PCR amplification can be achieved by using thermostable enzymes with proofreading exonuclease activities such as Vent (New England Biolabs) and Pfu (Stratagene) DNA polymerases.

Materials

- Taq DNA polymerase and dNTP set (Fermentas, KO163)
- 50 ng/ml target DNA
- 50 µM oligonucleotide primer incorporating the mutation
- 50 µM standard primer
- mineral oil
- chloroform

- thermocycler (LIAP, Ampligen PCR-3)   **Equipment**
- electrophoresis equipment
    for agarose gels (Pharmacia GNA100), for vertical PAA gels (Hoefer SE200), for electroelution (Pharmacia Extraphor 2014–001).
- Microfuge Eppendorf 5417C with aerosol-tight rotor type FA 45–30–11

- 10× Amplification buffer   **Buffers**
    1.34 M Tris-HCl, pH 8.5
    40 mM MgCl$_2$
    0.33 M (NH$_4$)$_2$SO$_4$
    0.2 M β-mercaptoethanol

- Gel-loading mix
  80 % glycerol
  0.01 % bromophenol blue
  0.01 % xylene cyanol
- PAAG electrophoresis buffer (10× TBE)
  1 M Tris pH 8.3
  0.83 M boric acid
  10 mM EDTA
- Agarose gel electrophoresis buffer
  40 mM Tris-acetate pH 8.2
  20 mM Na acetate
  1 mM EDTA

Procedure

Introduction of Modified Sequences at the Ends of Cloned DNA Fragments

1. Combine the following in a 0.5-ml microcentrifuge tube:
   5 µl 10× amplification buffer
   2.5 µl 4 mM 4dNTP mix
   1 µl (50 pmol) mutant primer
   1 µl (50 pmol) standard primer
   10 µl (500 ng) template DNA
   H_2O to 49.5 µl
   0.5 µl (2 U) Taq DNA polymerase
   mix and spin down in microcentrifuge; overlay reaction with
   50 µl of mineral oil.

2. Carry out PCR in an automated thermal cycler for 25 cycles
   under the following conditions:
   denaturation, 20 s, 95 °C
   hybridization, 20 s, at appropriate annealing temperature
   elongation, 20–60 s, 72 °C
   time of elongation depends on the length of amplified DNA.
   At last cycle extend elongation for an additional 7 min at
   72 °C.

Note. The amplified DNA fragment before restriction enzyme
digestion and cloning must be gel-purified to avoid template
DNA contamination in the PCR sample.

3. Add 50 µl of chloroform to the reaction, vortex and centrifuge briefly, remove the aqueous phase and add to it 5 µl of gel-loading buffer.

Note. After this step reaction mix in aqueous phase (blue) is used for direct preparative electrophoresis on agarose or poly-acrylamide gels and electroelution from gel slice.
Gel-purified mutant PCR product can be digested with restriction enzymes and used for cloning purposes.

Introduction of Mutations into the Middle of Cloned Sequence Using Overlap Mutant Primers

1. Combine the following in each of two 0.5 ml microcentrifuge tubes, adding oligonucleotides 1 and 2 to separate tubes:
   5 µl 10× amplification buffer
   2.5 µl 4 mM 4dNTP mix
   1 µl (50 pmol) mutant oligonucleotides 1 or 2
   1 µl (50 pmol) appropriate standard primer, forward or reverse
   10 µl (500 ng) template DNA
   H_2O to 49.5 µl
   0.5 µl (2 U) Taq DNA polymerase

   Synthesis by PCR

2. Mix by vortexing and centrifuge in a microcentrifuge to recover the contents.

3. Overlay reaction with 50 µl of mineral oil to prevent evaporation.

4. Carry out PCR in an automated thermal cycler for 30 cycles of denaturation, hybridization and elongation under the appropriate conditions.

5. At last cycle extend elongation for an additional 7 min at 72 °C.

Note. The amplified products of the two PCR reactions should be gel-purified, to avoid amplifying of template DNA in the second step PCR.

6. Transfer the aqueous phase to a fresh tube.

7. Add 50 µl of chloroform and 5 µl of gel-loading buffer to the reaction, vortex and centrifuge briefly.

Note. After this step reaction mix in aqueous phase (blue) is used for direct preparative electrophoresis on agarose gel and isolation from a gel slice.

8. After gel purification combine the following for the second PCR amplification:
   add in a 0.5 ml tube:
   10 µl (~100 ng) first amplified fragment
   10 µl (~100 ng) second amplified fragment
   5 µl 10× amplification buffer
   2.5 µl 4 mM 4dNTP mix
   1 µl (50 pmol) each flanking sequence primer, forward and reverse
   H_2O to 49.5 µl
   0.5 µl (2 U) Taq DNA polymerase

9. Transfer the lower aqueous phase carefully to a fresh tube and treat the sample with 1 vol of phenol/chloroform by vortexing for 20 s. Centrifuge at 14 000 rpm for 1 min.

10. Transfer the upper aqueous phase to a fresh tube and treat the sample with 1 vol of chloroform by vortexing for 20 s. Centrifuge at 14 000 rpm for 1 min.

11. Add 50 µl of chloroform to the reaction, vortex and centrifuge briefly.

12. Transfer the upper aqueous phase to a fresh tube and add 0.1 volume of 4 M NH_4 acetate and 3 vol of ethanol. Mix and leave at −20 °C for 30 min. Centrifuge at 14 000 rpm for 15 min.

13. Decant off the supernatant carefully, wash the pellet with 200 µl of 70 % ethanol by vortexing, centrifuge at 14 000 rpm for 5 min and remove the supernatant.

14. Dry under vacuum and dissolve in 20 µl of water.

15. Estimate the quantity of material recovered by running a sample in an agarose gel.
   After this step the mutant PCR product can be digested with restriction enzymes and used for cloning purposes.

Introduction of Mutations into the Middle of Cloned Sequence Using PCR-Created Megaprimer

1. Combine the following components in a 0.5-ml microcentrifuge tube for "megaprimer"-directed synthesis by asymmetric PCR reaction:
   5 µl 10× amplification buffer
   2.5 µl 4 mM 4dNTP mix
   1 µl (1 pmol) mutant oligonucleotide
   1 µl (50 pmol) appropriate standard primer, forward or reverse
   10 µl (500 ng) template DNA
   H_2O to 49.5 µl
   0.5 µl (2 U) Taq DNA polymerase

2. Mix by vortexing and centrifuge 2 s in a microcentrifuge to recover the contents.

3. Overlay reaction with 50 µl of mineral oil to prevent evaporation.

4. Carry out PCR in an automated thermal cycler for 30 cycles of denaturation, hybridization and elongation under the appropriate conditions. Extend the elongation at last cycle for an additional 7 min at 72 °C.

Note. It is necessary to remove standard primer from PCR product to avoid amplifying the template DNA by standard primers in the subsequent PCR step. This can be done on column (QIAgene PCR purification kit Cat. No. 28704) or by gel electrophoresis (see next two sections).

5. After purification of DNA combine the following components for the second PCR amplification in a 0.5 ml tube:
   5 µl 10× amplification buffer
   2.5 µl 4 mM 4dNTP mix
   10 µl (500 ng) template DNA
   30 µl (20 pmol) PCR created megaprimer
   1 µl (50 pmol) appropriate standard primer
   H_2O to 49.5 µl
   0.5 µl (2 U) Taq DNA polymerase

6. Mix by vortexing and centrifuge 2 s in a microcentrifuge to recover the content.

7. Overlay the reaction with 50 µl of mineral oil to prevent evaporation.

8. Carry out PCR in an automated thermal cycler for 30 cycles of denaturation, hybridization and elongation under the appropriate conditions.

Note. The amplified product should be gel-purified, to avoid presence of template DNA in the final PCR sample.

9. Transfer the aqueous phase to a fresh tube.

10. Add 50 µl of chloroform and 5 µl of gel-loading buffer to the reaction, vortex and centrifuge briefly.

Note. After this step reaction mix in aqueous phase (blue) is used for direct preparative electrophoresis on agarose or poly-acrylamide gels and for isolation from a gel slice.
Gel-purified mutant PCR product can be digested further with restriction enzymes and used for cloning purposes.

Purification of PCR Products by Polyacrylamide Gel Electrophoresis

1. Load sample onto the 8 % polyacrylamide gel and electrophorese.

Caution. Unpolymerized acrylamide is a potent neurotoxin. Handle with extreme care, and use gloves to protect hands.

2. After electrophoresis, stain the gel with ethidium bromide 0.5 µg/ml and visualize in UV light.

Caution. Ethidium bromide s a powerful mutagen. Use gloves to protect hands. Ultraviolet radiation is dangerous, particularly to the eyes. Protect eyes with UV filter shields

Note. Try to minimize the time of exposure of gel to UV light to prevent DNA damage, especially if the DNA is intended for cloning experiments.

3. Cut out the desired band carefully using a scalpel.
Elution of DNA from gel slice can carried out in Extrapor electroelution chamber as described in under "Materials".

Isolation of DNA from Agarose Gel by Centrifugation Method

1. Load the PCR sample onto the 1.5 % agarose gel containing 0.5 µg/ml of ethidium bromide and electrophorese at 75 V for 30–60 min. Use a wide slot on a large gel when large amounts of DNA are to be separated

2. After electrophoresis visualize DNA in UV light and make photograph of gel.

3. Cut out the target band carefully using a scalpel.

4. Transfer the agarose block containing the DNA to a 0.5-ml microfuge tube which has been punctured at its base and plugged with a small amount of glass wool. Place the tube into a 1.5-ml microfuge tube and centrifuge for 5 min at 10 000 rpm.

Note. After this step eluted DNA solution can be used in second PCR reaction. (see "Introduction of Mutations into the Middle of Cloned Sequence Using PCR-Created Megaprimer")

5. Remove the smaller tube, adjust the volume to 150 µl and treat the sample with 1 vol of phenol/H_2O by vortexing for 20 s to remove ethidium bromide and agarose. Centrifuge at 14 000 rpm for 1 min.

6. Transfer the upper aqueous phase carefully to a fresh tube and treat the sample with 1 vol of phenol/chloroform by vortexing for 20 s. Centrifuge at 14 000 rpm for 1 min.

7. Transfer the upper aqueous phase to a fresh tube and treat the sample with one volume of chloroform by vortexing for 20 s. Centrifuge at 14 000 rpm for 1 min.

8. Transfer the upper aqueous phase to a fresh tube and add 0.1 vol of 4 M ammonium acetate and 3 vol of ethanol. Mix and leave at −20 °C for 30 min. Centrifuge at 14 000 rpm for 15 min.

9. Decant off the supernatant carefully, wash the pellet with 200 µl of 70 % ethanol by vortexing. Centrifuge at 14 000 rpm for 5 min and pour away the supernatant.

10. Dry pellet under vacuum and dissolve in water.

11. Estimate the quantity of material recovered by running a sample in an agarose gel.

Preparation of Mutant PCR Fragments for Cloning

Both mutant PCR fragment and vector DNA should be digested with appropriate restriction enzymes to generate compatible ends for cloning. A sufficient amount of DNA for cloning can be obtained from a 50 µl volume of PCR reaction mixture.

1. Prepare the purified PCR fragment by digestion. The following 50 µl reaction is provided as an example:
   purified PCR fragment from 50 µl volume PCR mixture ($\sim$0.1 µg/ml)
   5 µl appropriate 10× restriction enzyme buffer
   5–25 U appropriate restriction enzyme.
   Add H_2O to 50 µl final volume.

2. Incubate at the appropriate temperature for 2–3 h.

3. If the restriction sites are sufficiently far from ends of PCR fragment, check that the reaction is complete by electrophoresis of a sample on a 1.5 % agarose gel.

4. Extract the DNA sample with one volume of phenol/chloroform.

5. Extract with chloroform.

6. Ethanol precipitate the DNA and resuspend in H_2O.

Ligation, Transformation and Selection of Mutant Clones

Ligation of digested vector and mutant PCR fragment and transformation can be done as described in Chapter 1 by Steinbergs and Tsimanis.

Selection for mutant clones:

Standard methods for selection (blue/white screening and/or appropriate resistance to antibiotics and phenotype of bacterial cells) of transformants can be used to identify clones which contain mutated sequences. If mutations create or remove a restriction site from the mutated DNA an appropriate restriction endonuclease can be used for screening the recombinant plasmids using standard methods of restriction analysis. Sequencing of mutant plasmids is preferred for checking the presence of mutation and the fidelity of amplification within a cloned fragment. This can be done by the dideoxy termination method using T7 DNA polymerase (see Chap. 2 by Jankevics)

Results

Three base substitutions in the central part of phage Qβ RNA replicase gene were made by PCR-directed mutagenesis using the method described (see "Introduction of Mutations into the Middle of Cloned Sequence Using PCR-Created Megaprimer"). Mutant primers used for each of the separate first-step reactions were:
GAAAATTATTTCGTCATTGTCAAGGATTC-5' mutant primer 2
5'-CTTTTAATAAAGCAGTTACTGTACCTAAGAACAGTAAG-3' template DNA
3'-GAAAATTATTTCGTCAATGACATGGATTCTTGTCATTC-5'
5'AGCAGTAACAGTTCCTAAGAACAGTAAG mutant primer 1
Proposed base changes in template DNA are underlined. Opposite to the mutant primers two standard primers were chosen so that they flanked the restriction sites for cloning needs. In the second PCR both previously obtained DNA fragments were mixed and amplified only with standard primers. The final mutant PCR product was digested with restriction enzymes BsiWI and ApaI and ligated in the previously digested recombinant plasmid which contains the complete genome of bacteriophage Qβ under control of phage T7 promoter. The presence of mutations in the plasmid was verified by DNA sequencing of DNA obtained from transformants.

References

Cormack B (1992) In: Ausubel F M et al. (eds) Current protocols in molecular biology. Wiley, New York, Unit 8.5

Higuchi R (1989) In: Erlich HA et al. (eds) PCR Technology. Stockton Press, pp 61–70

Ho SN, Hunt HD, Horton RM, Pullen JK, Pease LR (1989) Site directed mutagenesis by overlap extension using the polymerase chain reaction Gene 77:51–59

Nassal M, Rieger A (1990) PCR based site directed mutagenesis using primers with mismatched 3' ends. Nucl Acids Res 18:3077–3078

Perrin S, Gilliland G (1990) Site-directed mutagenesis using asymmetric polymerase chain reaction and a single mutant primer. Nucl Acids Res 18:7433–7438

Sharrocks AD, Shaw PE (1992) Improved primer design for PCR-based, site directed mutagenesis. Nucl Acids Res 20:1147

Tsai CH, Dreher TW (1993) In vitro transcription of RNAs with defined 3' termini from PCR-generated templates. BioTechniques 14:596–600

Vallette F, Mege E, Reiss A, Adesnik M (1989) Construction of mutant and chimeric genes using the polymerase chain reaction. Nucl Acids Res 17:723–733

Chapter 5

Analysis of Specific Protein-DNA Interactions

INDRIKIS MUIZNIEKS[*1] AND NILS ROSTOKS

Introduction

The central issue in the regulation of genome functions is the mechanism of sequence-specific protein-nucleic acid interactions. Gene expression, replication, recombination and DNA condensation in chromatin are steered by binding of regulatory protein ligands to specific sites in DNA. Numerous methods have been developed to study protein-DNA interactions. In this chapter we discuss two widely used and straightforward approaches to address this problem.

Electrophoretic mobility shift assay (EMSA), or gel retardation, or band shift assay characterize the capability of proteins to bind DNA fragments and to form the complexes which are stable in non-denaturing polyacrylamide gels and move slower than free DNA in electrophoresis.

DNA footprinting or DNA protection against the attack of degrading agents by the bound proteins allows the identification of the specific nucleotide sequences which are involved in binding. Additionally, both methods yield information about the structure of the complex and quantitative data about the kinetics of the interactions.

While countless modifications of the methods originally described have been published, appropriate adjustments are needed in every special case. Here, we would like to introduce our system that has worked well for the studies of the regulation of bacterial α-galactosidase (*rafA*) gene promoter P_{rafA}. Within 80 nucleotide base pairs (bp) P_{rafA} carries: (1) the binding site for RNA-polymerase; (2) the recognition sequence for the non-

[*] Corresponding author: Indrikis Muiznieks, phone: +371-7-322914; fax: +371-7-325657; e-mail: indrikis@acad.latnet.lv
[1] University of Latvia, Faculty of Biology, Kronvalda 4, LV1586 Riga, Latvia

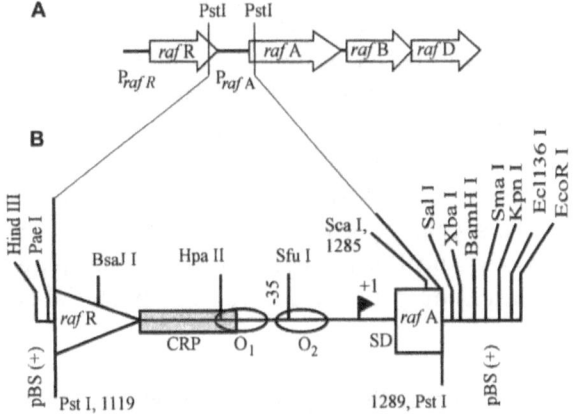

Fig. 5.1A–B. Structure of bacterial plasmid-borne *raf*-operon and P*rafA* promoter. **A** The *raf*-operon encodes functions required for the inducible uptake and utilization of raffinose in *Escherichia coli*. The expression of three structural genes is negatively controlled by the product of *rafR* gene, the RafR repressor. Homodimers of 36.8 kDa RafR bind to two operator sites O_1 and O_2, that flank the -35 sequence of the *raf* promoter P*rafA* (Aslanidis 1989, 1990). The *rafA* gene codes for α-galactosidase, a protein of estimated M_r 81.2 kDa, which is active in tetrameric form. Further members of the *raf*-operon are Raf permease (*rafB*) and sucrose hydrolase (*rafD*). The disaccharide melibiose is a natural inducer of the *raf*-operon. The expression of *raf*-genes is activated by cAMP receptor protein CRP in absence of glucose. The binding sequence for the CRP protein is immediately flanking the O_1 operator. As in the *lacZ* gene promoter, the centre of CRP binding site in P*rafA* is 61.5 bp 5' to the start point of mRNA synthesis. **B** The 170-bp *PstI* fragment, base pairs 1119 to 1289 of *raf* operon according to Aslanidis and Schmitt (1990), which carries the P*rafA* was cloned in polylinker of the phagemid pBS(+) creating plasmid pRU1330. Polylinker sites are indicated with *vertically rotated* names of the restriction enzymes. The polylinker sites were useful in creating differentially labelled DNA ends. RafR binding operator sequences O_1 and O_2 are depicted as *open ellipses*, the site for CRP protein binding is shown as a *shaded box* which partially overlaps with O_1. *Banner* shows the relative position of mRNA start. In the same experiment tandem dimer of the P*rafA* promoter fragment in *PstI* site of pBS(+) was obtained. Cutting the tandem dimer with restriction enzymes which are indicated within P*rafA* sequence generated permutated fragments of identical length, but with different positioning of protein-binding sites within the DNA molecule. The cloning in pBS(+) was intended also to obtain single-stranded DNA of P*rafA* for site-directed mutagenesis. The Amersham Sculptor in vitro mutagenesis system was used to create the mutations which abrogate binding capacity of either O_1 or O_2 (Muiznieks and Schmitt 1994). The sole *HpaII* site within P*rafA* is the product of mutagenesis. In P*rafA* derivatives which carry *HpaII* site the operator sequence O_1 has lost the ability to bind RafR, while CRP and O_2 sites are functional

specific activator, cAMP-receptor protein (CRP); (3) two binding sites for the cognate repressor (RafR). The structure of P_{rafA} is shown in Fig. 5.1. Protein binding at P_{rafA} reflects the basic principles of bacterial gene regulation. Meanwhile, the studies of multiple protein interactions within the condensed space of P_{rafA} provide novel insights for the characterization of the role of DNA topology in building protein-DNA complexes.

This chapter gives detailed protocols for making necessary protein and DNA preparations and carrying out the EMSA and DNase I footprinting analysis of a bacterial gene promoter, P_{rafA}.

Electrophoretic Mobility Shift Assay (EMSA)

Principle and Applications

EMSA was developed in the early 1980 s (Fried and Crothers 1981; Garner and Revzin 1981) and since then has undergone many modifications that allowed it to become a primary tool in a number of molecular biology applications (for reviews see Carey 1991; Lane 1992; Kerr 1995).

Advantages of this method are its relative simplicity, ability to resolve multiple protein-DNA complexes (Fig. 5.2) and the possibility to work with subpicomolar amounts of material. EMSA may be used for a variety of purposes such as: (1) analysing the ability of a DNA sequence to bind some protein factors; (2) finding an unknown protein factor that binds to a certain DNA sequence or vice versa; (3) studying structural and topological changes in protein or DNA caused by the molecular interactions; (4) exploring the thermodynamic and kinetic parameters of protein-DNA binding.

EMSA is usually performed in polyacrylamide gels (PAAG). Modifications of the method for agarose gels are described, but they are used mostly for observing DNA band shifts with very large protein complexes (Lieberman and Berk 1991). PAAG resolving power sets the limit for the length of DNA fragments used in EMSA. Considering the reduced electrophoretic mobility of protein-DNA complexes, the use of DNA fragments longer than 400–500 bp should be avoided. The lower limit of the DNA length for EMSA is within the oligonucleotide range. It is determined by the size of the specific protein-binding site on DNA, ca. 30 bp.

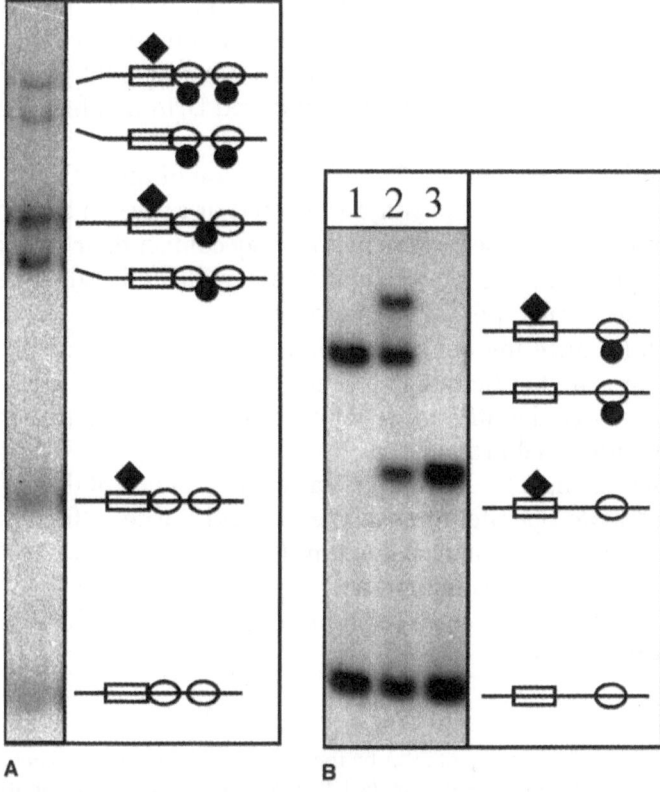

Fig. 5.2A–B. Analysis of protein-DNA complexes by EMSA. **A** The *wt* P*rafA* may form five different complexes with RafR and CRP. All types of the complexes can be resolved in 20-cm-long 4 % PAAG as exemplified by the autoradiogram on the *left* side of the panel. The structure of the complexes is explained on the *right*. RafR is shown as *filled circle* (·), CRP as a *filled diamond* (■). The *line* depicts the 221-bp *Eco*RI/*Hind*III fragment from pBS(+) which was isolated and labelled with [α-³²P]dATP and Klenow enzyme. *Open ellipses* stand for operator sites, the *box* for the CRP binding sites. If only one repressor dimer is binding to the DNA, no defined affinity for either operator is observed (Muiznieks and Schmitt 1994). Therefore in the pictogram the repressor is depicted in *intermediary position*. The distance between the centres of operator sites is 21 bp, exactly two turns of the helix in B-DNA. Both RafR dimers interact with the DNA from the same side of the molecule. The distance between the centres of CRP and RafR binding sites is 17–18 bp. CRP and RafR are mutually rotated at about 120–150° around the surface of the DNA tube. In the pictograms the proteins are depicted on the *opposite sides* of DNA. **B** EMSA with P*rafA* where O_1 has been inactivated by point mutation creating *Hpa*II restriction site. *Lane 1* 221-bp *Eco*RI/*Hind*III fragment bound only with RafR; *lane 2* with CRP and RafR; *lane 3* with CRP only. Calculations according to Eq. (1) give the value of ω ~ 1, consequently, there is neither interference nor cooperativity in binding of CRP and RafR to the mutated P*rafA*.The structures of protein-DNA complexes are explained in the *right side* of the panel

Specific applications of EMSA that have been important for our research are: (1) the determination of binding co-operativity of protein factors to an individual DNA fragment (Fig. 5.2); (2) the characterization of protein-induced DNA bending (Fig. 5.3).

Binding Co-operativity. The transcription of the majority of prokaryotic promoters is either repressed or activated by some protein factors. These protein factors may contact their binding sites independently or they may interact with each other and thus exhibit some hindrance or co-operativity at binding. In the case of hindrance the affinity of binding of each individual protein factor is higher than that of their joint binding. In the case of co-operativity the affinity of joint binding is higher than that for any individual protein.

An example for hindrance may be the binding of the *raf* repressor to two operator sites in the plasmid-borne raffinose operon promoter (Muiznieks and Schmitt 1994). Both operator sites are bound with the same affinity, however, if one of them is already bound by the repressor the other one is bound with ca. 13-times lower affinity.

Fig. 5.3A–C. Bending of DNA by the regulatory proteins. A Structure of the ▶ tandem dimer of the 170-bp *Pst*I fragment carrying *wt* and mutant P*rafA*. The restriction enzymes which cut only once per P*rafA* *Pst*I fragment monomer and were used to obtain permutated sequences are given *above* the figure. The permutated fragments were end-phosphorylated with PNK and [γ-^{32}P]ATP. For labelling, the *Pst*I fragment, which has a 5'-recessed end, was substituted by the *Sca*I fragment (blunt end). The distance between *Sca*I and *Pst*I sites in the 5'-part of the *rafA* gene is only 4 bp (Fig. 5.1). The localization of protein-binding sites within *Sca*I fragment from the P*rafA* dimer is nearly the same as in non-permutated sequence of *Pst*I fragment of P*rafA*. The relative localization of the protein-binding sites within the fragments is shown by the same *symbols* as in Fig. 5.2. The types of permutated fragments are denoted with *letters* on the *left margin* of the figure. B Bending of P*rafA* by RafR. EMSA was prepared with 0.2 ng of RafR and permutated P*rafA* fragments which are denoted *above* the lanes and in the *left margin* of Fig. 5.3A. The distance in bp between the centre of the binding site and the fragment midpoint is given *below* the lanes. The most pronounced mobility shifts are produced with those DNA fragments where the repressor binds near the midpoint of the fragment. C Bending of P*rafA* by CRP. EMSA was prepared with 0.12 ng of affinity-purified CRP and permutated P*rafA* fragments. The distance in bp between the centre of the binding site and the fragment midpoint is given *below* the lanes. The data show that both RafR and CRP bend P*rafA* DNA upon binding. Using Eq. (2) (see text), we have estimated that in the P*rafA* sequence the binding of CRP induces an angle of 85°, but the binding of one RafR repressor dimer an angle of 110°

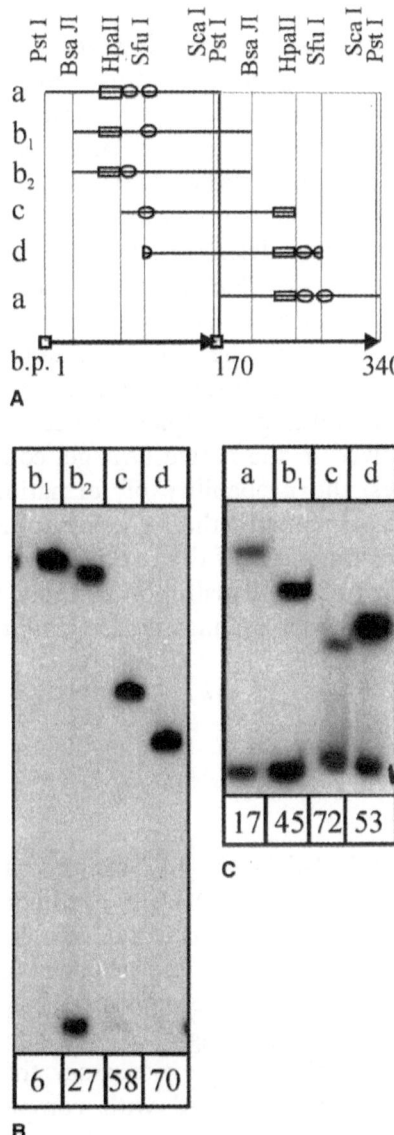

A

B

C

The co-operative binding of heterogeneous proteins is exemplified by the binding of CRP and *lac* repressor to their respective primary sites in the *lac* operon promoter (Hudson and Fried 1990; Vossen 1996). However, there are numerous papers on studies of protein binding co-operativity and protein-protein interactions employing the EMSA (see Pedersen 1992, Søgaard-Andersen and Valentin-Hansen 1993, Mao 1994, Kristensen 1996 and references therein). Two approaches have been developed to study binding co-operativity of protein factors: (1) protein distribution analysis (Fried and Crothers 1981); (2) binding competition assay. For detailed calculations and formulae see Hudson and Fried (1990) and Senear and Brenowitz (1991).

Both approaches characterize the relative protein-DNA binding constants through relative probabilities of formation of the corresponding complex. Since the protein-DNA complexes resolved in the native PAA gel contain large numbers of molecules, the probability of formation of each complex may be approximated by the frequency of its occurrence corresponding to the intensity of the band in autoradiography.

By protein distribution analysis, the co-operativity parameter ω_{P1P2} may be calculated according to the equation:

$$\frac{I_0 I_{P1P2}}{I_{P1} I_{P2}} = \omega_{P1P2} \tag{1}$$

Here, I_0, I_{P1}, I_{P2} and I_{P1P2} are the intensities of the autoradiography bands of free DNA; the Protein 1-DNA complex; the Protein 2-DNA complex and the double complex of both proteins with DNA, respectively. The proteins co-operate at binding DNA if ω_{P1P2} is >1, behave neutrally if ω_{P1P2} is ~ 1 and interfere with each others' binding if ω_{P1P2} is <1.

In binding competition assay, the multiple protein-DNA complexes are incubated with competing, non-specific DNA. To employ this method one of the protein factors must possess lower binding affinity for its site or, alternatively, the binding affinity of the same protein for another site must be different. Because one protein is more weakly bound than another, it is preferably transferred to competing DNA. In case of binding co-operativity this transfer is reduced in the presence of the second protein. To obtain the results one should compare the autoradiography band intensities in two binding assays: (1) with the both proteins; and (2) with the protein which is transferred to the competitor DNA more easily.

Protein-Induced DNA Bending. Protein-induced DNA bending plays an important role in building the spatial structure of the transcription complex. In EMSA, the protein binding-dependent mobility shift of DNA fragments is further increased by protein-induced DNA bending. Maximal mobility anomaly is observed when the bend is localized in the centre of the fragment and minimal when the bend is at the end of the fragment (Kolb 1983; Wu and Crothers 1984). In the "circular permutation assay" (Fig. 5.3), a tandem repeat of the DNA fragment containing a protein-binding site is cleaved with the restriction enzymes which cut only once per fragment monomer. Thus, a set of DNA fragments of the same length but with different protein-binding site location is obtained. The DNA fragments are complexed with the protein and run on native PAAG. The relative mobilities of the protein-DNA complexes are plotted against the position of the restriction sites in the DNA fragment (5→3'). The apex of the curve indicates the centre of the bend.

The relative bending angles can be calculated according to the equation:

$$\mu_M/\mu_E = \cos \alpha/2, \tag{2}$$

where μ_M is the mobility of the complex with protein bound at the centre of the DNA and μ_E = mobility of the complex with the protein bound at the end (Kim 1989; Thompson and Landy 1988).

Alternatively, DNA bending angles in protein-DNA complexes may be evaluated by comparing their electrophoretic mobility to a set of fragments which carry the standard DNA curvature elements, adenine tract determined bends (Zinkel and Crothers 1990).

DNA Footprinting

EMSA is helpful for characterizing the binding of specific proteins to DNA fragments which carry cognate recognition sites, but it does not provide sequence information about the structure of these sites. The analysis of protein-dependent DNA protection against non-processive degradation, DNA footprinting, is used to identify the sequences which are directly interacting with proteins. The principle of the method is depicted in Fig. 5.4.

Fig. 5.4A–D. Scheme of the footprinting experiment. **A** The components ▶ needed for DNA footprinting: singly end-labelled DNA fragment, DNA-binding protein, DNase I and chemicals for nucleotide base-specific degradation of DNA. Their preparation is described in Sections 5.1 to 5.3. **B** The protein-DNA binding reaction and partial degradation of unprotected part of the DNA fragment (Sect. 5.4). The protein-DNA complex is shown in the *right* part of the panel. The protein binding induces DNA bending. In the control reaction, the protein-unprotected fragment is subjected to DNase I degradation as shown in the *middle* part of the panel. To identify the DNA motif which is interacting with the protein, chemical sequencing reactions are carried out in parallel with the same fragment. Stochastic mixture of labelled and unlabelled DNA degradation products is generated. The concentration of DNA-degrading agents which are used in the reaction should produce one or less than one chain break per DNA molecule. Considering the huge number of molecules in the reaction (range of 10^9) this should result in statistically even distribution of degradation events over all the accessible sites for degradation. More than one attack of degrading agents per DNA molecule is depicted in picture just to make clear that numerous fragments are produced in the reaction. **C** The reaction is stopped, partially degraded DNA fragments are extracted, concentrated, denatured and electrophoresed in sequencing gel (see Sect. 5.4). Only the fragments which are produced from the labelled chain of the DNA will be visualized in the gel. The label from the other chain usually is removed by cutting away terminal 10–20 bases with an appropriate restriction enzyme. The small labelled fragment which is formed in this reaction does not produce interfering bands. The electrophoresis is carried out so that the bands shorter than 25–30 bp leave the gel. **D** Autoradiography and analysis of the footprint (Sect. 5.5). The analysis of DNA degradation patterns reveal: (1) regions of the protein-specific protection of DNA, the "windows", where the bands are missing due to the presence of DNA binding factor in footprinting reaction; (2) regions of DNA which are poorly cleaved by DNase I also in the absence of the protein due to some local structural features of the fragment, e.g. narrowed minor groove in oligo-T tracts; (3) DNase I hypersensitive sites which may be created by protein-DNA interactions usually as a consequence of protein-induced DNA bending and changing the configuration of the grooves. Poorly cleaved DNA regions are recognized in footprints as empty zones across all the lanes, both in the reactions with DNA binding proteins and in the controls. They may interfere with precise localization of the borders of the specific footprints. Minute structural modifications may influence the DNA sensitivity to DNase I degradation. The methylation of C residue in CG dinucleotide enhances the DNase I susceptibility of the neighbouring 5' phosphodiester bond, although the methyl-group of cytosine is not exposed in the minor groove of the DNA (Kochanek 1993)

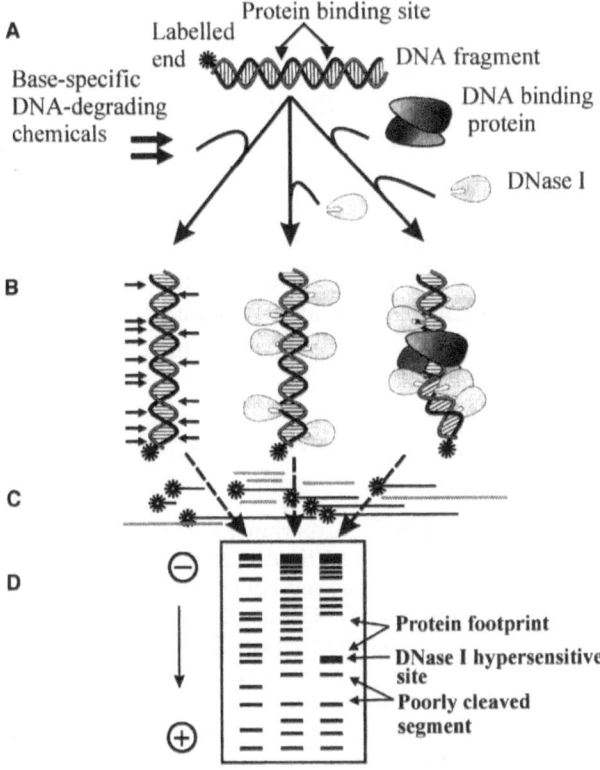

The method was proposed by D.J. Galas and A. Schmitz in 1978. The founder-fathers of DNA footprinting used DNase I as the DNA degrading tool. Since then a plenitude of degrading agents have proven their usefulness in DNA footprinting analysis. They may be classified into three main groups: (1) DNA degrading enzymes; (2) chemical and physical agents which produce free radicals; (3) chemical agents which modify nucleotides and prepare DNA for subsequent cleavage at the places of modification.

DNase I. High specific activity, stability at storage and reproducibility of the results obtained still make DNase I the enzyme of choice for the majority of DNA footprinting experiments (for recent review see Leblanc and Moss 1994). DNase I is an endonuclease which attacks DNA in the minor groove. The protein of 31 kDa molecular mass is active in the monomer form, and its structure is stabilized by Ca^{2+} ions (Lizarraga 1978). The degradation is non-processive. The enzyme cleaves preferentially after pyrimidine bases. Depending upon subtle modulations of DNA structure, e.g. minor alternations in the minor groove width or flexibility, DNase I will cleave some nucleotide motifs more easily or, on the other hand, less easily (Hogan 1989; Kochanek 1993). This can be considered as either a drawback of the method or, vice versa, can be employed to obtain additional information about the quest structures (Fig. 5.5). The molecular dimensions of DNase I are comparable to those of the proteins usually involved in DNA binding. This implies that the boundaries of the protected regions will be drawn with some extension. Some nucleotides immediately next to the bound proteins will remain inaccessible to the DNase I action.

Fig. 5.5. DNase I protection experiments (footprinting) of P_{rafA} with RafR ▶ and CRP proteins. The *EcoRI/HindIII* fragment from pRU1330 was end-phosphorylated with PNK and subsequently digested with *PaeI*. The experiments were performed with 0.5 ng of DNase I per reaction. Approximately one half of the DNA molecules was not cleaved by the nuclease. They build a *thick zone* on *top* of the gel. *Lanes 1* and *2* show chemical sequencing reactions with the P_{rafA} fragment using C- and G-specific modification reactions as denoted *below the lanes* (Maxam and Gilbert 1980). The *left margin* of the panel demonstrates the tracing of specific G and C pattern within the P_{rafA} sequence (Aslanidis and Schmitt 1990) which permits the precise localization of the protein-binding motifs. A number "1" *below* the figure denotes the presence of 10 ng of purified RafR or ca. 10 ng CRP from crude *E. coli* cell extract. "0" denotes the absence of the particular protein. *Lanes 3* and *9* are controls where the DNA was subjected to DNase I attack without protein protection. The regions of intrinsic resistance to DNase I attack are marked with *grey blocks* in the *right margin*. In *lanes 4* and *8* the DNase protection patterns with RafR and CRP, respectively, are shown. CRP produces pronounced bands of DNase I hypersensitivity in the centre of protected DNA segment, while RafR generates only minor bands of enhanced cleavage at the outer borders of the binding sequence. DNase I hypersensitive sites which are generated by the protein binding are marked in the *right margin* of the panel by *arrowheads*. EMSA (Fig. 5.3) has shown that both the proteins bend DNA. The differences in the DNase I hypersensitivity patterns imply that the manner of DNA bending by RafR and CRP is different. *Lane 7* demonstrates that both operators and CRP binding sequence of P_{rafA} can be occupied by the cognate proteins simultaneously. *Narrow white block* on the *right margin* of the panel shows the borders of CRP binding site. Two *broader white blocks* span the binding sequences of RafR. In *lanes 5* and *6* which are indexed by *M* the inducer of the *raf*-operon, melibiose, was added in the binding mixture to final concentration 10 mM. As expected the binding of RafR is weakened in the presence of the inducer (cf. *lanes 4* and *5*). The RafR-CRP-DNA complex is slightly more stable in the presence of melibiose than the RafR-DNA complex alone (*lanes 5* and *6*)

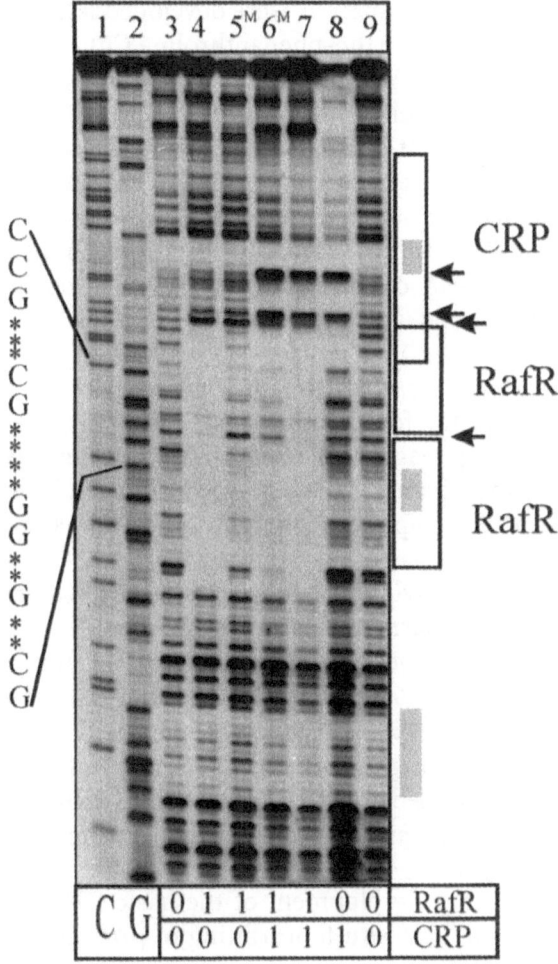

DNase II, Micrococcal Nuclease. DNase II and micrococcal nuclease are further enzymes which are employed for DNA footprinting. Micrococcal nuclease cuts almost exclusively at pA and pT bonds showing greater activity at $(A-T)_n$ than in homopolymeric runs of A and T (Fox and Waring 1987). These sequences are poorly cut by DNase I. The use of both of the enzymes provides mutually complementary data.

Hydroxyl Radical. The hydroxyl radical, generated by the reduction of hydrogen peroxide by iron(II), was first used to cut the DNA in footprinting assays by D.T. Tullius and B.A. Dombrovski in 1986. Numerous systems for the generation of free radicals for

DNA structure analysis were developed later: methydiumpropyl-EDTA.Fe(II), ortho-phenanthroline.CuI, photoinduction, etc. (Tullius 1991; Macgregor 1992; Baily and Waring 1995). In contrast to the bulky enzymes, the hydroxyl radicals can protrude into the closest boundaries of DNA and protein interactions, and they are less sequence specific than the nucleases, although they demonstrate some DNA secondary structure specificity. Some possibility still exists that unspecific denaturation of protein or protein-DNA complexes may take place during the time when the reactions which generate free radicals are initiated.

Dimethyl Sulphate. Dimethyl sulphate is the most widely used chemical for footprinting analysis among the nucleotide base-modifying reagents. The capability to penetrate through the cellular membranes makes it especially useful for in vivo genomic footprinting applications (Saluz and Jost 1993).

Others. Diethylpyrocarbonate, psoralen, osmium tetroxide and potassium permanganate can be considered as conformation-sensitive probes in DNA-degradation analysis (Runkel and Nordheim 1986; McCarthy and Rich 1991). In the absence of specific binding proteins these agents will preferably attack the regions of highly supercoiled, undertwisted DNA, partially single-stranded, melted DNA regions and four-way junctions. These agents are esspecially useful in the analysis of promoter structures.

A further development of the basic DNA-protection analysis technique is in situ footprinting of protein-DNA complexes following EMSA.

Quantitative analysis of protein-DNA interactions by means of DNA footprinting is a feasible, although seldom used approach (Rehfuss 1990).

▊ Materials

Equipment
- Devices for vertical PAAG electrophoresis, e.g., Minigel-Twin and Maxigel (Biometra)
- Device for horizontal agarose gel electrophoresis GNA100 (Pharmacia)
- UV transilluminator B89196 (Bioblock Sci.)
- Microcentrifuge, e.g., Beckman Microfuge E (Beckman Instruments, Inc.)

- Sorvall centrifuge RC-3B (Du Pont)
- French pressure cell press (American Instrument Co.)
- Ultra Turrax dispersing tool T25 (IKA Labortechnik)
- Beckman scintillation counter LS5800 (Beckman Instruments, Inc.)

- *Escherichia coli* DNA Polymerase I Large Fragment (Klenow **Materials**
  Fragment) (Boehringer Mannheim)

- DNase I, RNase-free or grade I (Boehringer Mannheim)
- T4 Polynucleotide Kinase (any available commercial vendor, e.g. New England Biolabs, MBI Fermentas)
- restriction enzymes (any available commercial vendor)
- shrimp alkaline phosphatase (Amersham-USB)
- 100 mM solutions of ultrapure dNTPs (Pharmacia)
- $[\alpha\text{-}^{32}P]$ dNTPs and $[\gamma\text{-}^{32}P]ATP$ (at 3000 Ci/mM, 10 mCi/ml) (Amersham)

Note. All the chemicals should be highest purity available!

- NucTrap Probe Purification Columns (Stratagene)
- plasmid DNA purification kits (Qiagen)
- X-ray films – Fuji RX
- intensifying screens for autoradiography – DuPoint Cronex Lightning Plus

- XL1-Blue Stratagene   **Bacterial**
  Genotype – *recA1 endA1 gyrA96 thi-1 hsdR17 supE44 relA1 lac*   **Strain**
  [F' *proAB lacI*^{q}ZΔM15 Tn*10* (Tet^r)]

- 2× TY medium   **Bacterial**
  16 g/l bacto-tryptone   **Growth**
  10 g/l yeast extract   **Medium**
  5 g/l NaCl

Note. Sterilize by autoclaving for 20 min at 121 °C

- Acrylamide stock for EMSA gels   **Solutions**
  39 % acrylamide
  1 % bis-acrylamide

Note. Store in dark bottle at 4 °C

Caution. Acrylamide is a neurotoxin and is readily absorbed through the skin. Always wear gloves when working with acrylamide and its solutions!

– Acrylamide stock for sequencing gels
  5.7 % acrylamide
  0.3 % bis-acrylamide
  7 M ultrapure urea
  10 % 10× TBE buffer

Note. Store in dark bottle at 4 °C
– Ammonium acetate
  10 M CH_3COONH_4, pH 8.0
– Ammonium persulphate (APS)
  10 % solution

Note. Store in aliquots at −20 °C as the APS is unstable in aqueous solutions. Use the aliquot only once

– Ca^{2+}/Mg^{2+} Solution
  5 mM $CaCl_2$
  10 mM $MgCl_2$
– cAMP
  10 mM Adenosine 3': 5'-Cyclic Monophosphate in TE Buffer

Note. Filter sterilize, store in aliquots at −20 °C

– Competitor DNA
  10 µg/ml of poly(dI):poly(dC) or poly(dA):poly(dT) in TE Buffer.

Note. Store in aliquots at −20 °C

– EDTA
  0.5 M ethylenediaminetetraacetic acid disodium salt dihydrate, pH 8.0

Note. EDTA will not dissolve until the pH 8.0 is set by the NaOH

– Ethanol
  100 % and 70 % (v/v) solutions
– Ethidium bromide
  10^{-4}% in 1× TAE

Note. Store in dark bottle at room temperature (20 °C)

Caution. Ethidium bromide is a powerful mutagen. Always wear gloves while handling gels or solutions containing the dye!

– Loading Solution I
  50 % glycerol
  0.05 % bromphenol blue

0.05 % xylene cyanol
- Loading Solution II
  0.5 M NaOH
  50 % formamide
  0.1 % bromphenol blue
  0.1 % xylene cyanol
- Melibiose
  100 mM solution in ddH$_2$O

Note. Filter sterilize

- PhenolCIA
  Phenol:chloroform:isoamyl alcohol (25:24:1)
  Store at 4 °C

Note. Phenol is preliminarily equilibrated with 0.5 M Tris-HCl and 0.5 M NaCl!

Caution. Phenol can cause severe burns. Always wear gloves when working with it. Any areas of skin that come into contact with phenol should be washed with soap and water. Do not use ethanol

- SDS
  10 % sodium dodecyl sulphate
- Sodium acetate
  3 M CH$_3$COONa, pH 4.8
- Stop solution
  0.5 % SDS
  0.13 M NaCl
  30 µg yeast tRNA per ml

- Buffer I   **Buffers**
  25 mM Tris-HCl, pH 8.0
  50 mM NaCl
  1 mM EDTA
  0.25 mg/ml lysozyme

Note. Prepare buffer without lysozyme and add it to the buffer just before use

- Buffer II
  2 % (v/v) Triton X-100
  40 mM Tris-HCl, pH 6.5
  0.5 M NaCl
  8 mM EDTA

- Buffer III
  100 mM Tris-HCl, pH 6.5
  20 mM EDTA
- Buffer D
  50 mM K phosphate, pH 7.5
  2 mM EDTA
  2 mM dithiothreitol (DTT)
  5 % (v/v) glycerol
- Buffer S
  10 mM Na phosphate, pH 6.8
  0.1 mM EDTA
  0.1 M NaCl
  50 % (v/v) glycerol
- Buffer W
  0.5 M K phosphate pH 7.5
  2 mM EDTA
  2 mM DTT
  5 % (v/v) glycerol
- DNA elution buffer
  0.5 M ammonium acetate
  10 mM magnesium acetate
  1 mM EDTA, pH 8.0
  0.1 % SDS
- DNase I stock and dilution buffer
  50 mM KCl
  50 mM Tris-HCl, pH 8.0
  1 mM DTT
  100 µg/ml bovine serum albumin
  50 % (v/v) glycerol
- Imidazole buffer, 10×
  0.5 M imidazole HCl, pH 6.4
  180 mM MgCl$_2$
  50 mM DTT
- Klenow enzyme buffer, 10×
  0.5 M Tris-HCl, pH 7.6
  0.1 M MgCl$_2$
- Polynucleotide Kinase buffer, 10×
  0.5 M Tris-HCl, pH 7.6
  0.1 M MgCl$_2$
  50 mM DTT
  1 mM spermidine
  1 mM EDTA, pH 8.0
- Protein-DNA binding buffer, 10×

100 mM Tris-HCl, pH 8.0
100 mM KCl
10 mM EDTA
10 mM DTT
0.5 mg/ml BSA
30 % (v/v) glycerol
0.1 % Nonidet P-40

Note. If the binding of CRP to the DNA is to be tested, include 2 mM cAMP in the 10× binding buffer

- TAE electrophoresis buffer, 50×
  2 M Tris-acetate, pH 8.0
  0.05 M EDTA
- TBE electrophoresis buffer, 10×
  0.89 M Tris-Borate, pH 8.0
  0.02 M EDTA
- TE buffer
  10 mM Tris-HCl, pH 8.0
  1 mM EDTA
- TE 0.1 buffer
  10 mM Tris-HCl, pH 8.0
  0.1 mM EDTA

5.1
Isolation of DNA Fragments

Procedure

A variety of methods for purifying DNA fragments exist either from PAAG or agarose gel. In our hands the best results have been obtained with the modification of the method which is described by Sambrook (1989).

Basic Protocol

1. Set up the incubation of 5 μg of Qiagen column-purified plasmid DNA which contains P_{rafA} with appropriate restriction enzyme(s) in 50 μl total volume of the reaction mixture.

2. While the plasmid DNA is digested, prepare a 5 % PAAG. In a measuring cylinder mix:
   4.0 ml acrylamide stock for EMSA, 0.64 ml 50× TAE, 0.25 ml 10 % APS, ddH$_2$O to 32 ml.
   Add 30 μl of TEMED and mix well. Pour a 120×120×1.2 mm gel

with 25-mm-wide slots, pre run 0.5 h with 15 V/cm at room temperature (20 °C).

3. Add 5 µl of Loading Solution I to the digested DNA probes, mix, and load them into the gel slots. Continue the electrophoresis until the bromphenol blue marker has migrated 2/3 of the gel length.

4. Stain the gel with ethidium bromide and visualize the DNA fragments in reflected UV light. Cut out the band of interest.

5. Transfer the gel slice into a 1.5-ml Eppendorf tube and add 400 µl of DNA elution buffer.

6. Incubate the tube overnight at 37 °C, if possible on the rotator.

7. Spin the tube briefly in microcentrifuge to collect the condensation from tube walls and collect the elution buffer carefully trying not to transfer PAAG pieces.

8. Elute the DNA fragment once more with half the volume of elution buffer for a couple of hours. Collect the elution buffer as previously and combine with the first one.

9. Precipitate the DNA with isopropanol (1.0 vol) or 100 % ethanol (2.5 vol) and collect the DNA pellet by centrifugation in micro-centrifuge for 15 min at maximal speed.

10. Discard the supernatant, wash the pellet with 70 % ethanol and centrifuge for 5 min.

11. Dry the pellet at 37 °C and dissolve in 200 µl of ddH$_2$O. Add 25 µl of 3 M sodium acetate, pH 4.8, precipitate, wash and dry as previously.

12. Dissolve the DNA pellet in water or TE 0.1 buffer.

The procedure should yield ca. 0.2 µg (1.5 pM) of 170–210 bp fragments. Run a 1/10 aliquot of the fragment preparation in 1.5 % agarose gel and determine the DNA concentration by comparing the ethidium bromide fluorescence of the fragment with the fluorescence of the equal length bands containing 10 and 20 ng DNA.

Options Good quality DNA can be recovered also from agarose gels by using electroelution or by centrifugation of agarose gel slice through glass wool.

Numerous reagent kits for purification of DNA fragments from agarose gels are commercially available, e.g., QIAquick Gel Extraction Kit from Qiagen or Sephaglas BandPrep Kit from Pharmacia.

In some cases, for EMSA, but not for footprinting assays, one can omit isolation of the DNA fragment and proceed directly with DNA labelling, if there are no other DNA fragments in the restriction hydrolysate that might be confused with protein-DNA complex(es).

It is also possible to label the DNA fragments first, directly in the restriction mixture and then purify labelled fragments from agarose or PAAG. DNA bands in the gel can be located by autoradiography and purified by any of above-mentioned procedures.

5.2
Preparation of Protein Factors for EMSA

EMSA is based on specific protein-DNA interactions often simulating the in vivo conditions when only several protein molecules (~10 as in *lac* repressor case) specifically bind to a single DNA site per genome. Consequently, not only highly purified proteins, but also crude cell extracts may be used in EMSA. The gene cloning approaches allow one to obtain recombinant bacteria which produce high amounts of the proteins of interest. This allows one to use dilute crude extracts in binding reactions, thereby preventing the interference of endogenous nucleases and proteases. Binding buffers without magnesium ions restrict the activity of most nucleases.

Procedure

The repressor of the plasmid-borne bacterial raffinose catabolism operon is encoded by the *raf*R gene (Aslanidis and Schmitt 1990) which has been subcloned into high copy number plasmid pUC8 under the control of the *lacZ* promoter. Repressor protein expression is induced upon addition of IPTG and protein is accumulated in the form of inclusion bodies that are easily prepared, purified and reconstituted in the active form (Aslanidis 1990).

RafR

1. Transform the *Escherichia coli* strain XL1-Blue, with the plasmid pRU984 (Aslanidis 1990).

2. Transfer a single colony into 300 ml of 2× TY medium supplemented with ampicillin (100 µg/ml) and IPTG (1 mM) and grow the culture overnight at 37 °C with shaking.

3. Harvest the cells by centrifugation for 30 min at 5000 rpm in the Sorvall centrifuge, discard the supernatant, resuspend the cells in 5 ml of Buffer I and incubate for 30 min on ice.

4. Disrupt the cells by three passages through the French press. Allow the lysate to cool down on ice between the passages.

5. Sediment the inclusion bodies by centrifugation for 15 min at 10 000 rpm, 4 °C and discard the supernatant.

6. Resuspend the inclusion bodies in 5 ml of Buffer II, homogenize using an Ultra Turrax dispersing tool T25 three times for 20 s and keep the homogenate on ice for 30 min.

Note. An alternative to the Ultra Turrax dispersing tool may be sonication three times for 20 s at 22 kHz, however, very strong foaming is observed due to presence of Triton X-100.

7. Repeat step 5.

8. Resuspend the inclusion bodies in 5 ml of Buffer III, homogenize three times with Ultra Turrax and keep 30 min on ice. Repeat the centrifugation/homogenization step three times.

9. Resuspend the inclusion bodies in 0.5 ml of Buffer I without the lysozyme. Add SDS to the final concentration of 0.1 % and allow the solubilization of the RafR protein to proceed overnight at 4 °C.

10. Pellet the non-soluble inclusion bodies by centrifugation in the micro-centrifuge for 10 min at maximal speed, dispense the supernatant containing the functional repressor in 20 µl aliquots and store at −70 °C.

Protein preparation is analysed using 10 % PAA-SDS gel as described in Sambrook 1989. Average concentration of protein preparation is about 2 µg/µl as judged by SDS-PAGE and the preparation contains only minor contaminants (see Fig. 2B in Aslanidis 1990).

The gene for CRP is cloned and constitutively overexpressed in a high copy-number plasmid pBG2 (Breul 1993).

CRP

1. Transform the *E. coli* strain XL1-Blue with the plasmid pBG2.

2. Transfer a single colony into 300 ml of 2× TY medium supplemented with ampicillin (100 µg/ml) and grow the culture overnight at 37 °C with shaking.

Basic Protocol

3. Harvest the cells by centrifugation for 30 min at 5000 rpm in the Sorvall centrifuge, discard the supernatant, resuspend the cells in Buffer I to give $OD_{600} = 100$, incubate for 30 min on ice, freeze at −70 °C and thaw on ice.

4. Disrupt the cells by three passages through the French press. Allow the lysate to cool down on ice between the passages.

5. Centrifuge the lysate for 15 min at 10 000 rpm, 4 °C and discard the pellet containing cell debris. Aliquot the supernatant and keep at −70 °C.

Chromatography Through the cAMP-Agarose

1. Dialyse 3 ml of the supernatant from the step 5 overnight against two changes of 500 ml of Buffer D. Centrifuge the dialysate at 4 °C, 30 min, 15 000 rpm to sediment the precipitated protein.

Option

2. Make a 5 ml column of cAMP-agarose (Pharmacia), pre-wash it with 5 column volumes of Buffer W +1 M NaCl and equilibrate with 10 vol of buffer D at flow rate 12 ml/h.

3. Load the dialysate to the column at flow rate 3.5 ml/h and wash with 10 column volumes of Buffer W (12 ml/h).

4. Elute CRP with Buffer W +2 mM cAMP at flow rate 2 ml/h. Collect 0.5 ml fractions and check the protein concentration. CRP begins to elute at the end of the first column volume and leaves the column in ½ of its volume.

5. Dialyse the peak fractions of CRP overnight at 4 °C against 100 vol of Buffer S, aliquot and store at −70 °C.

In our hands, the use of more than 1000-fold diluted crude CRP extracts and affinity-purified CRP preparations gave similar results in EMSA and footprinting.

5.3
Labelling of DNA Fragments for EMSA and Footprinting

Although many non-radioactive DNA labelling methods exist today, radioactive labelling is still a widely used, fast and convenient method ensuring the highest quality and sensitivity for various molecular biology applications. The particles emitted by the decaying radioactive isotope penetrate the photographic film, collide with silver halide crystals and generate precipitates of silver atoms. The isotope of choice to prepare radioactively labelled DNA fragments for EMSA is ^{32}P because its β-emission energy is much stronger than that of other often used isotopes (1.709 MeV compared to 0.167 MeV for ^{35}S). Particles emitted by ^{35}S can penetrate the film emulsion only to a depth of 0.25 mm which is not enough when wet gel is covered with Saran Wrap. In contrast, ^{32}P generates β particles which penetrate water or plastic to a depth of 6 mm and pass completely through an X-ray film. This allows us to take autoradiograms of wet gels covered with Saran Wrap as well as make use of intensifying screens that enhance the image ca. fivefold.

Several approaches have been developed to produce labelled DNA fragments with high specific activity but not all of them are suitable for EMSA or footprinting. Random priming method yields DNA probe with specific activity $>10^9$ cpm/μg, however, the label is spread within the fragment and the length of the fragments is not uniform. We recommend two approaches that may be used to generate labelled DNA fragments for EMSA with a specific activity $>10^7$ cpm/μg DNA: (1) labelling 3'-recessed ends with *E. coli* DNA Polymerase I Klenow fragment and appropriate [α-^{32}P]dNTP; and (2) labelling dephosphorylated 5'-ends with T4 polynucleotide kinase and [γ-^{32}P]ATP. The choice of method depends on the ends of the DNA fragment produced by different restriction enzymes. Only recessed 3'-ends are labelled by the Klenow enzyme. T4 polynucleotide kinase may be used to label protruding 5'-ends as well as blunt ends. Recessed 5'-ends are labelled with low efficiency. In this case, the use of imidazole buffer instead of the standard kinase buffer may improve the efficiency of labelling.

▨ Procedure

Labelling with the Klenow Enzyme

The Klenow enzyme adds complementary deoxynucleotides to the hydroxyl groups at the recessed 3' ends of the DNA fragment. In contrast to *E. coli* DNA Polymerase I the Klenow enzyme possesses only 5' to 3' polymerase and 3' to 5' exonuclease activities. If one of the deoxynucleotides in the reaction is substituted by its [α-^{32}P] analogue, the reaction product will be a DNA fragment with one or both ends labelled depending on the restriction enzyme(s) used and on the labelled deoxynucleotide included in the reaction. In our experiments we labelled both ends of P$_{rafA}$ carrying the *EcoRI/Hind*III fragment from the plasmid pRU1330 (Fig. 5.1) with [α-^{32}P]dATP.

1. Mix in the Eppendorf tube following reaction components:
   up to 1.5 pM of DNA fragment,
   2.5 µl of 10× Klenow Buffer,
   10 µCi (~3.3 pM) of [α-^{32}P]dATP,
   1 µl of mix of other dNTPs (2 mM each) to fill the ends of DNA fragment,
   2 U of Klenow enzyme (labelling grade),
   ddH$_2$O to 25 µl.

2. Incubate for 30 min at 37 °C.

3. Increase the reaction volume to 50 µl with ddH$_2$O and add 1 µl of 5 M NaCl. Extract twice with PhenolCIA.

4. Increase the volume of the aqueous phase to 75 µl and add 25 µl of 10 M ammonium acetate. Precipitate with 250 µl of 100 % ethanol.

5. Spin for 10 min in microcentrifuge, wash the pellet with 70 % ethanol, air-dry.

6. Repeat steps 4 and 5. Control the radioactivity of supernatant with Geiger counter. Two ethanol precipitation/washing steps in presence of 2.5 M ammonium acetate remove more than 95 % of unincorporated label.

7. Perform the Cherenkov counting of the dry DNA pellet. Normally, the specific activity of the sample should be 2–5×10^7 cpm/µg.

8. Dissolve the sample in TE 0.1 buffer to make specific activity 1×10^5 cpm/µl.

Note. To avoid the loss of pellet during ethanol precipitation, place tubes in centrifuge with cap hinge at the top and note the position of pellet. Draw off ethanol carefully so as not to disturb the pellet.

Option   A frequently employed method to remove the unincorporated label from the reaction mix is gel filtration through Sephadex G-50 columns (Sambrook 1989). Instead of self-made columns it is recommended to use commercially available NucTrap Probe Purification Columns from Stratagene and follow the manufacturer's instructions.

Labelling with T4 Polynucleotide Kinase

T4 polynucleotide kinase (PNK) catalyses the transfer of the γ-phosphate group of ATP to a 5'-OH terminus of the DNA; therefore it is possible to label DNA using $[\gamma\text{-}^{32}P]ATP$. PNK can catalyze either the forward reaction, namely transfer of a phosphate to a 5'-OH group, or drive the exchange reaction, causing the transfer of the terminal 5'-phosphate group of DNA to ADP and afterwards the rephosphorylation of DNA by transfer of labelled γ-phosphate to DNA. ADP must be in excess amount. Here, we describe only the forward reaction for which the DNA must first be dephosphorylated. To calculate the concentrations of the termini of nucleic acid molecules to be labelled use Table 5.1.

Dephospho-   There are several enzymes that catalyse cleavage of 5'-phosphate
rylation   groups from DNA fragments leaving hydroxyl groups necessary for PNK. Those are bacterial alkaline phosphatase (BAP), calf intestinal alkaline phosphatase (CAP) and shrimp alkaline phos-

Table 5.1. Size-concentration relationship of linear double-stranded DNA

Size of double-stranded DNA (in base pairs)	Amount of DNA required to contribute 1 pM of 5' termini (in µg)
50	1.7×10^{-2}
100	3.3×10^{-2}
250	8.4×10^{-2}
500	1.7×10^{-1}

phatase (SAP). All of them carry out the same reaction, however, CAP and SAP have an essential advantage over BAP – they can be completely inactivated by heating, whereas BAP can be inactivated only by multiple phenol/chloroform extractions.

1. Mix the following components in an Eppendorf tube: DNA fragment to be dephosphorylated (1.5 pM), 5 µl of 10× SAP buffer (supplied with SAP), 1.0 U of SAP (0.2 U catalyze removal of phosphate from 1 pM of DNA ends), ddH$_2$O to 50 µl.

2. Incubate for 1 h at 37 °C.

3. Heat the reaction mix for 15 min at 65 °C, then increase the volume till 100 µl and add 2 µl of 5 M NaCl.

4. Extract twice with PhenolCIA.

5. Precipitate the aqueous phase with 2.5 vol of 100 % ethanol for 20 min on ice.

6. Spin in a micro-centrifuge for 10 min at maximal speed, wash the pellet with cold 70 % ethanol and air dry.

1. Add to the dephosphorylated dry DNA (up to 1.5 pM of 5' **Labelling** ends; see Table 5.1 to calculate concentration of your DNA fragment) the following components: 2 µl of 10× PNK buffer, 10 µCi (~3.3 pM) of [γ-^{32}P]ATP, 20–30 U of PNK, ddH$_2$O to 20 µl.

2. Incubate for 30 min at 37 °C.

3. Increase the volume of reaction mix to 100 µl, add 2 µl 5 M NaCl and extract once with PhenolCIA.

4. Separate the DNA from unincorporated [γ-^{32}P]ATP as described above.

5. Perform the Cherenkov counting of the dry DNA pellet. Normally, the specific activity of the sample should be >2–5×10^7 cpm/µg.

6. Dissolve the sample in TE 0.1 buffer to specific activity 1×10^5 cpm/µl.

Note. Ammonium ions are inhibitors of PNK therefore the DNA fragment after preparation from PAAG according to Sambrook (1989) must be carefully purified. DNA molecules with blunt or

5'-recessed ends are labelled less efficiently than those with 3'-recessed ends therefore it is recommended to increase the amount of PNK to 30–40 units. To obtain the effective phosphorylation of 5'-recessed ends it is also recommended to use imidazole buffer and include in the reaction mixture polyethylene glycol (PEG 8000) in concentrations ranging between 4 and 10 %.

Depending on the number of protein-DNA complexes formed in EMSA, good bands on X-ray film will be produced by 500–1000 cpm of the labelled fragment. To obtain the sufficiently strong signal in autoradiography it is advisable to take for one EMSA reaction the amount of labelled DNA corresponding to 500–1000 cpm, depending on the number of bands expected.

Preparation of the Fragment for Footprinting

Footprinting experiments require a *singly end-labelled* DNA fragment. To obtain this by phosphorylation with PNK, the labelled fragment must be additionally digested with a restriction enzyme which releases a small portion from one end. This procedure with the P_{rafA}-carrying plasmid pRU1330 is facilitated by the flanking sites from pBS(+) polylinker (Fig. 5.1). The fragment was cloned into the *Pst*I site, the *Eco*RI/*Hind*III fragment was isolated, labelled with PNK + [γ-^{32}P]ATP and digested with *Ecl*136II or *Pae*I to remove the label from one end of the fragment.

1. The following reaction was assembled in the microcentrifuge tube:
   20 µl of phosphorylated DNA fragment from above, 5 µl of 10× appropriate restriction buffer, 40 U of *Pae*I or *Ecl*136II, ddH$_2$O to 50 µl

2. Incubate at 37 °C for 1 h. It is important to remove one end of the fragment completely. This is ensured by excess enzyme in the reaction.

3. Extract twice with PhenolCIA.

4. Precipitate and wash the DNA with ethanol. Air dry and count the radioactivity.

5. Dissolve in TE 0.1 to final specific activity 1×10^5 cpm/µl. Store at 4 °C.

For one footprinting reaction ca. 30 000 cpm of labelled DNA are needed.

5.4
Protein-DNA Binding Reactions

Protein-DNA complexes are formed by mixing stoichiometric amounts of the DNA fragment and active protein. If the specific activity of the DNA is in the range of 3×10^7 cpm/µg, as little as 0.1–1.0 ng or several femtomoles of fragment are used in the reaction.

The protein concentration in the reaction depends on the quality of protein preparation and on the number of its binding sites on the DNA fragment. Usually not all protein molecules have retained their binding ability after purification. If the DNA carries more than one binding site, the amount of the cognate protein must be increased accordingly. In most cases, sub-saturating protein concentrations when the band of free DNA is still visible in the gel are optimal for the interpretation of EMSA. The right protein amount for binding reactions can be found only empirically by titration of labelled DNA fragments with series of dilutions of the protein preparation.

Efficiency of the formation of protein-DNA complexes is strongly influenced by the composition of binding buffer. In our work addition of DTT, BSA and 0.1 % Nonidet P-40 to the binding reaction favoured protein-DNA interactions. Glycerol in the binding buffer not only had a positive effect on protein binding but also allowed direct loading of the incubations' mix on PAAG without much mixing/pipetting. On the other hand, addition of Mg^{2+} ions in the binding buffer had no positive effect, moreover, it caused degradation of the DNA fragment especially when crude cell lysates with CRP were used. It is not possible to give any general recommendations concerning the buffer composition, mostly because of different binding conditions for different proteins (see Hassanain 1993).

In our hands the protein-binding buffer composition which is given above has worked well both with RafR and CRP, in EMSA and in footprinting assays. However, we would like to note that even for the very well studied CRP-*lac* promoter interactions binding conditions described by different authors may vary. Binding of particular protein factors to their sites demands that special ligands be added, e.g., specific binding of CRP occurs only in the presence of cAMP. Studies of bacterial repressor-operator interactions may require the addition of specific inducers that cause the dissociation of the complex. cAMP must be included not only in the binding buffer ($200 \mu M$) but also in the electrophoresis buffer ($20 \mu M$ end concentration).

All the nucleic acid binding proteins exhibit some degree of non-specific affinity. This can cause smearing of retarded bands or even appearance of non-specific protein-DNA complexes. This is especially important when crude cell lysates are used. Unspecific binding to the quest DNA may be avoided by increasing the salt concentration in the reaction mixture. A more frequently used approach to overcome this problem is adding of "non-specific" competitor DNA. For this purpose sonicated fish sperm or calf thymus DNA is often employed, although these DNAs may carry sequences which mimick specific binding sites. The use of synthetic competitor DNAs, e.g., poly(dA):poly(dT) or poly(dI):poly(dC), is preferable.

Formation of the protein-DNA complexes is usually carried out at room temperature, however, this may vary with the purpose of the experiment. Incubations at $4\,°C$ or at $37\,°C$ are described also.

The time for complex formation is usually chosen between 5 and 30 min. However, when binding of several consecutively added proteins is investigated reaction times may be increased up to 1 h to allow the binding to reach equilibrium.

All these variations in procedures just point out the necessity of empirical determination of individual reaction conditions for different protein factors and DNA sequences.

We describe here the protocol for the binding reaction which worked well with the proteins involved in the regulation of the *raf*-operon. Similar conditions were applicable also for the binding of crude and affinity-purified human transcription factor AP-2 to the cognate DNA sequences.

Procedure

Binding Reaction in EMSA

1. Mix in the Eppendorf tube:
   3000 cpm or ca. 1 fM of labelled DNA fragment, an appropriate amount of the protein, 1 µl of 10× protein-binding buffer,

cofactors (cAMP, 200 µM; melibiose 10 µM – 1 mM), 1 µg of competitor DNA [poly(dA):poly(dT)]
ddH$_2$O to 10 µl.

2. Incubate binding reactions for 20 min at room temperature (~20 °C).

3. Load the binding reactions directly on the non-denaturing PAAG without addition of dyes (this is possible due to the presence of 3 % glycerol in the binding reactions) and load 1 µl of Loading Buffer I in the side lanes of the gel to control the migration of the samples in the gel.

   In control reactions DNA fragments without proteins or without cofactors are incubated. Additional control is provided by the reaction without competitor DNA. In most reactions with purified RafR and CRP the competitor DNA was omitted since its presence did not influence the binding of the proteins to P$_{rafA}$.

   Typical amounts of the proteins used in our work were about 1 ng of purified RafR and CRP, 1:5000 dilution of the crude extract of CRP over-producing bacterial cells which were disrupted in French Press at OD$_{600}$ = 100 (Figs. 5.2–5.3).

   Specific binding of CRP to DNA is observed only in the presence of cAMP. Melibiose is the natural inducer of the *raf*-operon. This disaccharide abrogates RafR binding and it was used to study the protein-protein interactions at P$_{rafA}$.

Binding and Footprinting Reaction

1. Mix in the Eppendorf tube:
   30 000 cpm or ca. 10 fM of labelled DNA fragment, appropriate amount of the protein, 5 µl of 10× protein-binding buffer, cofactors (cAMP, 200 µM; melibiose 10 µM – 1 mM)
   ddH$_2$O to 50 µl.

2. Incubate binding reactions for 20 min at room temperature (~20 °C).

3. While the incubation is in progress, heat the Stop Solution to 37 °C and mix well.

4. Treat each reaction identically in the following manipulations.

Note. Process no more than three samples simultaneously to achieve similar results. Add 50 µl of Ca^{2+}/Mg^{2+} Solution and incubate at room temperature for 1 min. Add 3 µl of appropriate DNase I dilution, mix gently, but thoroughly, and incubate at room temperature for 1 min.

5. Terminate the reaction by adding 100 µl of Stop Solution. Mix well.

6. Extract the reaction with 200 µl PhenolCIA.

7. Transfer the upper, aqueous phase to a fresh tube, add 2 µl of 3 M Na-acetate and 500 µl of 100 % ethanol. Precipitate on ice for 20 min.

8. Spin down the DNA in microcentrifuge at maximal speed for 10 min. Carefully remove the supernatant, wash with 70 % ethanol, and air dry.

9. Resuspend the pellet in 4 µl Loading Solution II by vortexing and flicking the tube. Heat at 95 °C for 2 min and chill on ice for at least 2 min.

10. Load onto a 6 % polyacrylamide sequencing gel. Run the gel at 1200–1500 V in 1× TBE buffer until the bromphenol blue is at the bottom of the gel.

The optimal amount of DNase I for the footprinting reaction is to be determined empirically in pilot experiments. The typical DNase I concentrations added per reaction vary between 0.2–1.0 ng. This should allow approximately one random nick per labelled DNA molecule. The dilutions of the grade I DNase I are made in 50 % glycerol-containing buffer from 1 mg/ml stock solution. The dilutions and the stock solution may be stored at least 6 months at −20 °C.

To save the chemicals, the preliminary footprinting reactions can be performed in 1/3 of the described scale with no binding proteins added to the reaction and analysed on 120×200×0.5 mm denaturing PAA gels.

In control reactions DNA fragments are incubated without binding proteins or without cofactors which are needed for binding.

Unlike EMSA, in footprinting assay the DNA should be completely saturated with the binding proteins. For this purpose we used 10 ng of purified RafR and CAP per footprinting reaction (Fig. 5.5). When competitor DNA was added in footprinting with crude CRP extracts, the DNase I concentration per reaction was increased.

To localize the protein-binding sites, the footprinting samples are run in the sequencing gel in parallel with base-specific chemical degradation products of the same DNA fragment (Maxam and Gilbert 1980).

5.5
Electrophoretic Analysis

Procedure

Electrophoresis in EMSA

Dimensions of the PAAG for EMSA depend mostly on the number of expected shifted bands and on the molecular weight of the complexes. If only a couple of low molecular weight bands are formed, it is possible to use short gels (ca. 10 cm). However, if several large protein-DNA complexes are expected and, especially, if they have similar sizes, it is advisable to make longer gels (20 cm or more).

Long gels are also recommended when EMSA is used to assess topological features of DNA which are induced upon binding of protein factor, e.g., protein-induced DNA bending.

The standard PAAG thickness in EMSA is 1 mm. Thinner gels are more easily dried after electrophoresis, but the size of the wells may become too small. Gels which are 2–2.5 mm thick may be used when it is essential to load enough cell lysate, e.g., to detect DNA bound protein by Western blot.

Four to 5 % PAAG are most frequently used in EMSA. Such gels have a pore diameter ca. 16–20 nm and provide sufficient frictional force on protein-DNA complexes to resolve them according to their molecular mass and/or structural peculiarities, e.g., bent DNA structures.

PAAG pore diameter depends also on the degree of polymer cross-linking. Instead of the standard 29:1 or 38:2 acrylamide to bis-acrylamide ratio, EMSA gels usually have lower cross-linking at acrylamide to bis-acrylamide ratios of 39:1 or 75:1. In our experiments we used PAA gels with acrylamide to bis ratio 39:1. This is sufficient to separate on a 20 cm long gel all the protein-DNA complexes formed by RafR and CRP with P_{rafA}.

Our experience shows that electrophoresis in 1× TAE buffer gives sharper bands and better resolution than in 0.5× TBE. To avoid buffer exhaustion during prolonged runs, an electrophore-

Table 5.2. Migration of marker dyes in the native PAA gels

Percentage of gel	Bromphenol blue[a]	Xylene cyanol[a]
3.5	100	460
5.0	65	260
8.0	45	160
12.0	20	70
20.0	12	45

[a] The numbers are the approximate sizes of DNA fragments (in nucleotide pairs) with which the dyes would comigrate.

sis chamber with buffer recircularization should be used. Electrophoresis is usually carried out at the same or lower temperature as binding reactions. At least a 0.5 h long pre-run is recommended to guarantee even distribution of ligands in the buffer and gel and to allow stabilization of the current.

Electrophoresis is carried out at ca. 10 V/cm until the samples have migrated appropriate distances. To observe the progression of the electrophoresis, dye markers are added in the lanes next to the binding reactions (see Table 5.2). When the samples have migrated the desired distance in the gel, the current is stopped, the glass plates are disassembled and the gel is either vacuum-dried or covered with Saran Wrap and directly subjected to auto-radiography.

Electrophoresis of the Footprints

The footprinting analysis is performed in denaturing sequencing gels and include all the usual steps for processing these gels.

The length of the gel run depends on the DNA fragment size and the localization of the protein-binding site within the fragment. Fragments of 100–600 bp may be used in footprinting, with the protein-binding site not closer than 30 bp from the labelled end. Protein-binding sites as far as 400 bp from the labelled end can be used, but require longer electrophoresis, and the bands are not so sharp.

Good results can be obtained with any type of sequencing gel. To ensure a uniform load of radioactivity in every slot it is recommended to count every probe before dissolving and to adjust the volume of the Loading Solution II according to the amount of cpm in the tube. The optimal width of the slots is 6–8 mm, with gel thickness at the top ~0.2 mm.

Gel Autoradiography

The standard method for the detection of radioactively labelled nucleic acids is autoradiography, although recently phosphor-imaging systems have become available allowing direct scanning of gels and blots without use of X-ray films. Even when employing phosphor-imaging systems, it is advisable to make also an autoradiogram of the gel for your record.

Both wet and dried gels can be subjected to autoradiography. Autoradiography of wet gels covered with Saran Wrap is preferred when further manipulations with the gel are planned, e.g., when localized bands are cut out and radioactivity is counted or when localized protein-DNA complexes are excised and their footprints are made. Dry gels, however, provide higher sensitivity and better quality pictures.

Autoradiography of EMSA gels is usually carried out with two intensifying screens to shorten the exposure times which depend on distribution of labelled DNA in bands and which may vary from several hours to several days.

When quantitative experiments are carried out and densitometry of exposed films is planned it may be reasonable to use pre-flashed X-ray films for autoradiography. Films are pre-exposed to a short (<1 ms) flash of light that activates the silver halide crystals in the emulsion. Crystals in these films have a prolonged linear response to emitted β particles and fluorescent light of the intensifying screens.

Sequencing gels should be dried before autoradiography. With gels which contain more than 20 000 cpm per slot we have not noticed significant differences between autoradiographs done at room temperature or at $-70\,°C$.

References

Aslanidis C, Schmid K, Schmitt R (1989) Nucleotide sequence and operon structure of plasmid-borne genes mediating uptake and utilization of raffinose in *Escherichia coli*. J Bacteriol 171:6753–6761

Aslanidis C, Muiznieks I, Schmitt R (1990) Successive binding of *raf* repressor to adjacent *raf* operator sites in vitro. Mol Gen Genet 223:297–304

Aslanidis C, Schmitt R (1990) Regulatory elements of the raffinose operon: nucleotide sequence of operator and repressor genes. J Bacteriol 172:2178–2180

Baily C, Waring MJ (1995) Comparison of different footprinting methodologies for detecting binding sites for small ligand on DNA. J Biomol Struct Dyn 12:869–898

Breul A, Assmann H, Golz R, von Wilcken-Bergmann B, Müller-Hill B (1993) Mutants with substitutions for Glu-171 in the catabolite activation protein (CAP) of *Escherichia coli* activate transcription from the *lac* promoter. Mol Gen Genet 238:155–160

Carey J (1991) Gel retardation. In: Sauer RT (ed) Protein-DNA Interactions. Methods Enzymol. vol 208. Academic Press, London, pp 103–117

Fried M, Crothers DM (1981) Equilibria and kinetics of *lac* repressor-operator interactions by polyacrylamide gel electrophoresis. Nucleic Acids Res 9:6505–6525

Galas DJ, Schmitz A (1978) DNAse footprinting: a simple method for the detection of protein- DNA binding specificity. Nucleic Acids Res 5:3157–3170

Garner MM, Revzin A (1981) A gel electrophoresis method for quantifying the binding of proteins to specific DNA regions: application to components of the *Escherichia coli* lactose operon regulatory system. Nucleic Acids Res 9:3047–3060

Hassanain HH, Dai W, Gupta SL (1993) Enhanced gel mobility shift assay for DNA-binding factors. Anal Biochem 213:162–167

Hogan ME, Robertson MW, Austin RH (1989) DNA flexibility variation may dominate DNase I cleavage. Proc Natl Acad Sci USA 86:9273–9277

Hudson JM, Fried MG (1990) Co-operative interactions between the catabolite gene activator protein and the *lac* repressor at the lactose promoter. J Mol Biol 214:381–396

Kerr LD (1995) Electrophoretic mobility shift assay. Methods Enzymol 254:619–632

Kim J, Zwieb C, Wu C, Adhya S (1989) Bending of DNA by gene-regulatory proteins: construction and use of a DNA bending vector. Gene 85:15–21

Kochanek S, Renz D, Doerfler W (1993) Differences in the accessibility of methylated and unmethylated DNA to DNase I. Nucleic Acids Res 21:5843–5845

Kolb A, Spassky A., Chapon C, Blazy B, Buc H (1983) On the different binding affinities of CRP at the *lac, gal* and *mal*T regions. Nucleic Acids Res 11:7833–7852

Kristensen H-H, Valentin-Hansen P, Søgaard-Andersen L (1996) CytR/cAMP-CRP nucleoprotein formation in *E. coli*: the CytR repressor binds its operator as a stable dimer in a ternary complex with cAMP-CRP. J Mol Biol 260:113–119

Lane D, Prentki P, Chandler M (1992) Use of gel retardation to analyze protein-nucleic acid interactions. Microbiol Rev 56:509–528

Lieberman PM,. Berk AJ (1991) The Zta *trans*-activator protein stabilizes TFIID association with promoter DNA by direct protein-protein interaction. Genes Dev 5:2441–2454

Lizzaraga B, Sanchez-Romero D, Gil A, Melgar E (1978) The role of Ca^{2+} on pH-induced hydrodynamic changes of bovine pancreatic deoxyribonuclease A. J Biol Chem 253:3191–3195

Mao C, Carlson NG, Little JW (1994) Cooperative DNA-protein interactions. Effects of changing the spacing between adjacent binding sites. J Mol Biol 235:532–544

McCarthy JG, Rich A (1991) Detection of an usual distortion in A tract DNA using $KMnO_4$: effect of temperature and dystamycin on the altered conformation. Nucleic Acids Res 19:3421–3429

Macgregor RB (1992) Photogeneration of hydroxyl radicals for footprinting. Anal Biochem 204:324–327

Maxam AM, Gilbert W (1980) Sequencing end-labelled DNA with base-specific chemical cleavages. In: Grossman L, Moldave K (eds) Nucleic acids. Methods Enzymol, vol 65. Academic Press, London, pp 499–599

Muiznieks I, Schmitt R (1994) Role of two operators in regulating the plasmid borne *raf* operon of *Escherichia coli*. Mol Gen Genet 242:90–99

Pedersen H, Søgaard-Andersen L, Holst B, Gerlach P, Bremer E, Valentin-Hansen P (1992) cAMP-CRP activator complex and the CytR repressor protein bind co-operatively to the *cytRP* promoter in *Escherichia coli* and CytR antagonizes the cAMP-CRP-induced DNA bend. J Mol Biol 227:396–406

Rehfuss R, Goidisman J, Dabrowiak JC (1990) Quantitative footprinting analysis. Binding to a single site. Biochemistry 29:777–781

Runkel L, Nordheim A (1986) Chemical footprinting of the interaction between left-handed Z-DNA and anti-Z-DNA antibodies by diethylpyrocarbonate carbethoxylation. J Mol Biol 189: 487–501

Saluz HP, Jost JP (1993) Approaches to characterize protein-DNA interactions in vivo. Crit Rev Eukaryot Gene Expr 3:1–29

Sambrook J, Fritch EF, Maniatis T (1989) Molecular cloning: a laboratory manual, 2nd edn. Cold Spring Harbor Laboratory, Cold Spring Harbor, New York

Senear DF, Brenowitz M (1991) Determination of binding constants for cooperative site-specific protein-DNA interactions using gel mobility shift assay. J Biol Chem 266:13661–13671

Søgaard-Andersen L, Valentin-Hansen P (1993) Protein-protein interactions in gene regulation: the cAMP-CRP complex sets the specificity of a second DNA-binding protein, the CytR repressor. Cell 75:557–566

Thompson JF, Landy A (1988) Empirical estimation of protein-induced DNA bending angles: applications to 1 site-specific recombination complexes. Nucleic Acids Res 20:9687–9705

Tullius TD (1991) DNA footprinting with the hydroxyl radical. Free Radic Res Commun 12–13 Pt 2:521–529

Tullius TD, Dombrovski BA (1986) Hydroxyl radical "footprinting": high resolution information about DNA-protein contacts and application to lambda repressor and Cro protein. Proc Natl Acad Sci USA 83:5469–5473

Vossen KM, Stickle DF, Fried MG (1996) The mechanism of CAP-*lac* repressor binding cooperativity at the *E. coli* lactose promoter. J Mol Biol 255:44–54

Wu H-M, Crothers DM (1984) The locus of sequence-directed and protein-induced DNA bending. Nature 308:509–513

Zinkel S, Crothers DM (1990) Comparative gel electrophoresis measurement of the DNA bend angle induced by the catabolite activator protein. Biopolymers 29:29–38

Chapter 6

Gel Electrophoresis, Transfer and Hybridization of Nucleic Acids

Illar Pata[*1] and Erkki Truve[2]

Introduction

Gel electrophoresis is the basic tool for size fractionation of nucleic acids. The method has unparalleled resolving power and is performed with considerable ease and speed. In an electric field nucleic acids move through the pores in the gel towards the anode according to their molecular weight. After electrophoresis, the location of nucleic acids within the gel can be directly visualized in ultraviolet light by a dye ethidium bromide. If desired, these bands may be excised from the gel, nucleic acids purified and used in various manipulations (e.g. cloning, labeling). Furthermore, it is possible to identify a particular sequence in a composite population of nucleic acids by means of a blotting technique. The nucleic acids are separated by gel electrophoresis and then transferred from the gel to a filter so that their relative positions are preserved. The fragments of interest are identified by hybridization with a complementary nucleic acid probe.

Principles and Applications

This nucleic acid separation and hybridization technique was originally described by E. Southern (1975), and in his honor has been termed Southern blotting for analysis of DNA and Northern blotting for analysis of RNA. This chapter intends to give complete protocols for both of these techniques. Moreover, each step (gel electrophoresis, transfer, hybridization and autoradiography) can be followed and performed independently. The main differences between methods used for DNA and RNA work exist in the electrophoresis protocols. The transfer and hybridization

[*] Corresponding author: Illar Pata, phone: +372–7–420201; fax: +372–7–420286; e-mail: ipata@ebc.ee

[1] University of Tartu, EE 2400 Riia 23, Tartu, Estonia

[2] Institute of Chemical Physics & Biophysics, Akadeemia tee 23, EE 0026 Tallinn, Estonia

procedures are quite similar for DNA and RNA, and are therefore depicted in detail only in the DNA subchapter.

We concentrate on agarose gel electrophoresis in this chapter. Polyacrylamide gels are used for separating small nucleic acid fragments (up to 0.5 kb) and are characterized in Chapter 3 by Avota and Licis. Agarose gels allowing resolution between 0.2 and 25 kb are used in standard Southern and Northern analysis techniques.

Analysis of DNA by Southern Blotting

Southern blotting is a technique for identifying a particular DNA sequence in a heterogeneous population of DNA molecules (Southern 1975). The DNA fragments are resolved by gel electrophoresis, denatured in situ and transferred from the gel to a membrane filter. The DNA immobilized on the filter is hybridized with the labeled probe and locations of the DNA bands complementary to the probe are determined. Southern blotting can be used to analyze sequences in cloned DNA (plasmids, cosmids etc.) as well as in genomic DNA. Genomic DNA is extremely complex and requires a certain degree of care. We will give protocols that enable the performance of genomic DNA analysis. For studying cloned DNA, the protocols may be greatly simplified (see Notes and Comments).

Analysis of RNA by Northern Blotting

Northern blotting is a technique where RNA molecules are separated by gel electrophoresis and detected by hybridization of a labeled probe on the membrane. The method enables the determination of the presence, abundance and size of the specific message.

Nondenaturing gels used for the separation of DNA molecules can be used also for the separation of RNA. However, as single-stranded RNA molecules often possess strong secondary structure elements, they do not always migrate in a nondenaturing gel according to their length. Therefore denaturing agents like formaldehyde (Lehrach et al. 1977), glyoxal (McMaster and Carmichael 1977) or methyl mercuric hydroxide (Bailey and Davidson 1976) are added to the gels. In addition, RNA samples must be denatured prior to loading the gel in the denaturation buffer at

65 °C. Highly expressed messages can be detected by Northern blotting in both total RNA as well as in poly(A)$^+$ mRNA, but for rare messages the usage of poly(A)$^+$ fraction of RNA is preferable.

Materials

Equipment
- Horizontal gel electrophoresis apparatus (Hoefer HE 33 Minnie Submarine Unit)
- DC power supply (EPS 600, Pharmacia)
- Gel casting platform
- Gel tray and combs
- Microwave oven (commercially available oven with power ≥800 W)
- Water bath (Hoefer Water Bath PR840)
- Hybridization oven and bottles (Mini10 from Hybaid)
- Hand-held radioactivity monitor (Berthold LB 1202)
- Autoradiography cassettes (Kodak BioMax Cassette with Bio-Max MS Intensifying Screen)
- Film developing system (Kodak X-Omat M43A) or alternatively phosphoimager (e.g. Molecular Dynamics Phosphor-Imager SI)
- Pre-flash unit (Sensitize, Amersham; optional)

Materials
- Saran Wrap
- Parafilm (American Can Company)
- Polaroid film (type 665 or 667)
- Nylon filter (e.g. Hybond-N or Hybond-N+, Amersham)
- 0.45 μm syringe filter (SFCA filter, Nalgene)
- X-ray film (Hyperfilm MP from Amersham)

Reagents
- Ethidium bromide: 10 mg/ml in water (**Caution.** Ethidium bromide is a powerful mutagen and suspected carcinogen, wear gloves!)
- Bacteriophage λ DNA, cut with HindIII
- Restriction endonucleases (from Fermentas, Pharmacia, Promega etc.)
- 20 mg/ml bovine serum albumin
- 5 M NaCl
- Nonlabeled deoxynucleotides: dATP, dGTP, dTTP, 2.5 mM of each
- Labeled deoxynucleotide: [α-^{32}P]dCTP (3000 Ci/mmol)

- 10 mg/ml sonicated salmon sperm DNA in deionized water
- Diethyl pyrocarbonate (**Caution.** Carcinogen, work under fume hood!)
- 0.5 M NaOH
- RNA molecular weight markers (from Promega etc.)

Electrophoresis Buffer (TAE or TBE)

- 1× TAE
  40 mM Tris-acetate pH 8.0
  1 mM EDTA
- 0.5× TBE
  0.045 mM Tris-borate pH 8.3
  1 mM EDTA
- 10× sample buffer
  0.25 % bromophenol blue
  0.25 % xylene cyanol
  50 % glycerol (v/v)
- TE buffer
  10 mM Tris-HCl pH 8.0
  1 mM EDTA
- Denaturation solution
  0.5 M NaOH
  1.5 M NaCl
- Neutralization solution
  0.5 M Tris-HCl pH 7.5
  1.5 M NaCl
- Transfer buffer (10× SSC)
  1.5 M NaCl
  0.6 M Na·citrate pH 7.0
- OLB solution A
  1.25 M Tris-HCl, pH 8.0
  0.125 M $MgCl_2$
  1.8 % (v/v) β-mercapthoethanol
- OLB solution B
  2 M Hepes (N-[2-hydroxyethyl]piperazine-N'-[2-ethanesulfonic acid]), pH 6.6
- OLB solution C
  random hexadeoxyribonucleotides, suspended in TE buffer at 90 ODU_{260} per ml
- OLB buffer - mix together solutions A:B:C in ratio of 100:250:150. Store at −20 °C

Solutions and Buffers

– 50× Denhardt
  1 % polyvinylpyrrolidone,
  1 % Ficoll 400,
  1 % bovine serum albumin in deionized water

Note. Filter-sterilize the solution.

– Hybridization solution (without formamide)
  6× SSC
  5× Denhardt
  0.5 % SDS
  100 µg/ml salmon sperm DNA

Note. Denature salmon sperm DNA before adding at 95 °C for
5–10 min followed by rapid chilling on ice.

– Filter wash solutions
  2× SSC, 0.5 % SDS (low stringency)
  0.2× SSC, 0.1 % SDS (medium stringency)
  0.1× SSC, 0.1 % SDS (high stringency)
– X-ray film developing solution
  0.35 % (w/v) metol
  6 % (w/v) Na_2SO_3
  0.9 % (w/v) hydroquinone
  4 % (w/v) Na_2CO_3
  0.35 % KBr
– X-ray film fixation solution
  28 % $Na_2S_2O_3 \cdot 5H_2O$
– 10× MOPS buffer
  0.2 M MOPS (3-[N-morpholino]propanesulfonic acid)
  0.05 M Na·acetate pH 7.0
  0.01 M EDTA

Note. 10× MOPS buffer should be autoclaved. Do not worry that
the solution turns yellow.

– RNA loading dye
  mix together 100 µl 10× MOPS buffer
  350 µl concentrated formaldehyde
  1 ml formamide (deionized)
  200 µl 10× loading dye
  20 µl 10 mg/ml ethidium bromide
– Depurination solution (optional)
  0.2 M HCl

6.1
Electrophoresis of DNA Using Agarose Gels

Procedure

Agarose gels are used for separation and purification of DNA fragments from 0.2 to 50 kb in length. Several factors (agarose percentage, buffer, applied voltage, amount of loaded DNA) must be taken into consideration for obtaining desired separation results in the shortest possible time. Choose the *optimal agarose* concentration for resolving DNA fragments of interest (Table 6.1). Gels with agarose percentage below 0.5 % are troublesome to handle, and over 2 % difficult to dissolve and pour. Use special brands of agarose (e.g. NuSieve, FMC BioProducts) for casting gels with high agarose percentage. The commonly used *electrophoresis buffers* are Tris-acetate (TAE) and Tris-borate (TBE). At a given agarose percentage, TAE resolves better the larger and TBE the smaller DNA fragments. TAE is preferred in analysis of supercoiled DNA (e. g. plasmid minipreps) while TBE is used in extended electrophoretic runs due to its better buffering capacity. The *applied voltage* has an effect on the separation results of DNA fragments larger than 5 kb. Although smaller DNA fragments can be run at high voltages (10–15 V/cm, measured as distance between electrodes) reducing the electrophoresis time, fragments larger than 5 kb get compressed. These must be resolved at low voltages (1–2 V/cm). The *tracking dyes* incorporated into the sample loading buffer (usually bromophenol blue and xylene cyanol) serve as convenient markers to monitor the progress of the electrophoretic run. The migration rate of the dyes depends primarily on the agarose concentration of the gel. In a standard 1 % agarose/TAE gel bromophenol blue comigrates with DNA fragments of 0.5 kb and xylene cyanol with 5 kb. In gels with lower (higher) agarose percentage the dyes comigrate with larger (smaller) DNA fragments, respectively (e.g. in 0.5 %

Standard Protocol

Table 6.1. Optimal agarose gel concentrations for resolution of linear DNA

Agarose percentage (%)	0.4	0.6	0.8	1.0	1.2	1.4	2
Range of separation (kb)	2–50	1–20	0.8–12	0.5–8	0.3–6	0.2–4	0.1–2

TAE bromophenol blue travels at 1 kb, and in 2 % TAE at 150 bp). The fluorescent dye *ethidium bromide* is used to visualize DNA in gels. For routine purposes we recommend to add ethidium bromide directly to the gel solution (addition of ethidium bromide to buffer in the electrophoresis tank is not required). However, while performing extended electrophoretic runs, it is better to run the gel without the dye and stain the gel afterwards. The *amount of DNA* loaded into the slot depends on the complexity of the sample and the proportion of the fragment to be analyzed within that sample. It should be noted that a single DNA band of 20–100 ng is properly visible (using a slot 0.5 cm wide), but even as little as 2 ng can be detected by additional staining and de-staining procedures. On the other hand, a DNA band of more than 200 ng results in band smearing and trailing. Therefore, restriction digests are loaded typically in amounts of 0.1–0.5 µg for plasmid analysis, 2–4 µg for cosmids and bacteriophage λ, and 10–30 µg for genomic DNA.

Preparation of the Gel

1. Seal the gel casting tray with the tape if it is opened at the ends. Set the tray horizontally onto the bench or platform. Position the gel comb so that about 1 mm remains between the bottom of the wells and the base of the tray. This ensures that the wells are formed and will not be broken when the comb is withdrawn.

   Note. If casting low percentage (below 0.5 %) or low melting agarose gels, set the comb higher.

2. Prepare a sufficient amount of electrophoresis buffer to fill the electrophoresis tank and to prepare a gel. The amount of gel solution required is determined both by the dimensions of the tray and thickness of the gel (which is usually 0.5 cm).

3. To prepare a gel, measure a volume of electrophoresis buffer into a flask and add the desired amount of agarose. Use a flask whose volume is at least three times the volume of the solution. Heat the suspension in a microwave oven using several 20–30 s intervals until the agarose is completely dissolved (no lumps visible). Swirl the flask gently in between the intervals to ensure even mixing. If the volume of solution has decreased due to evaporation, add deionized water to obtain initial volume and bring to a boil.

4. Cool the solution by swirling the flask in 50–60 °C water bath and add, if desired ethidium bromide to a final concentration of 0.5 µg/ml. Mix, avoiding creating air bubbles.

5. Pour the gel solution into the gel tray. Let it harden for at least 20 min.

Note. The low percentage and low melting agarose gels should be cooled for further 20 min in the fridge to gain hardness.

6. After the gel is completely set, carefully withdraw the comb and remove the tape. Mount the gel in the electrophoresis tank. Add electrophoresis buffer until the gel is covered with 2–3 mm of buffer.

7. Prepare the DNA samples by mixing them with 1/10 vol of 10× sample buffer. This can be done conveniently on a piece of Parafilm. Load the samples into the slots. It is advisable to load DNA molecular weight markers to one slot which helps to estimate the sizes of unknown DNAs.

1. Check that the electric leads have been attached so that DNA will migrate into the gel towards the anode (positive lead). Set the voltage to the desired level (1–15 V/cm, measured as distance between the electrodes). Run the gel until the bromophenol blue and xylene cyanol dyes have migrated to the distance which gives satisfactory results of separation of DNA fragments. **Running the Gel**

2. Turn off the power supply. If ethidium bromide was incorporated in the gel, DNA can be directly visualized in ultraviolet light. Otherwise, stain the gel by soaking it 20 min in electrophoresis buffer or water containing 0.5 µg/ml ethidium bromide. If minor quantities of DNA (below 5 ng) are detected, reduce the background fluorescence of the gel by destaining it in water for additional 10–20 min.

3. Photograph the gel using an UV light source, the appropriate photocamera equipped with the red filter and Polaroid film (type 665 or 667). Alternatively, the results can be recorded with the aid of a videocamera.

It is very convenient to use small agarose gels for routine everyday purposes, if the sizes of analyzed DNA fragments are approximately known (e.g. plasmid minipreps, PCR products, restriction controls etc.). The solutions with different agarose concentration can be prepared (normally with ethidium bro- **Minigels** mide) and stored at 65 °C for at least 1 week. A precast gel may be stored at room temperature, if upon hardening slots are filled with electrophoresis buffer or water and the gel is wrapped in Saran wrap. Minigels are usually run at high voltages (5–20 V/cm) and completed in 20–30 min. The buffer in the tank needs to be replaced only when it gets exhausted during runs.

Restriction Digestion of Mammalian Genomic DNA for Southern Blot Analysis

Support Protocol   Restriction endonucleases cleave *high molecular weight DNA* relatively slowly. Therefore the reaction is usually carried out overnight. For optimal results, it is important to ensure the completeness of digestion.

1. Digest 10–20 μg of genomic DNA (>50 kb) with appropriate restriction enzyme(s). Carry out the digestion in 400 μl of total volume containing an optimal digestion buffer, 200 μg/ml BSA, using fourfold excess of restriction enzyme at the appropriate temperature. Mix the reaction components thoroughly but gently.

Note. The quality of genomic DNA preparation is crucial. Before starting the experiment, it is advised to perform a test reaction (with 0.2 μg of genomic DNA in 20 μl reaction volume for 1 h) to determine whether the DNA can be efficiently digested. Include a control (DNA in reaction buffer with no restriction enzyme added) to assure that DNA will not be nonspecifically degraded in the presence of Mg^{2+}. Analyze the samples on agarose gel.

2. After 4 h, analyze a 10 μl aliquot of digestion on 0.8 % agarose gel. Run it together with the same amount of undigested DNA and judge for the completeness of digestion.

3. If the digestion is not complete (considerable amount of the DNA sample migrates at the same rate as the undigested control), add another aliquot of restriction enzyme (20–80 U) and continue incubation overnight.

4. If the digestion is complete, add 1/10 of 5 M NaCl and precipitate DNA by 2.5 vol of ethanol in a microfuge. Carefully remove the ethanol and dissolve the precipitate in 20–40 μl TE.

Note. Do not overdry the precipitate, otherwise it will be difficult to dissolve.

5. Cast a gel (usually 0.8 % agarose in 0.5× TBE) as described under "Preparation of the Gel" (without ethidium bromide), and add 1/10 vol 10× sample loading buffer to DNA samples. Be sure to include DNA molecular weight markers (e.g. λ DNA *Hind*III digest). Before loading onto the gel, incubate the DNA samples at 56 °C for 5 min.

Note. Prior to use, wash the electrophoresis apparatus and the gel casting tray thoroughly. Avoid contamination of DNA samples.

6. Electrophorese the gel at 1–2 V/cm until desired separation is achieved (usually 18 h). Stain and photograph the gel and proceed with transfer protocol (Sect. 6.2).

6.2
Capillary Transfer of DNA to a Nylon Filter

Transfer of DNA fragments from the electrophoresis gel to a filter support can be achieved by different means. Capillary transfer (Southern 1975) is the most widely used technique, in which the DNA fragments are carried from the gel onto a filter by upward *capillary force*. This is achieved by placing dry paper towels on top that absorb liquid through the gel. Alternative methods to capillary transfer are vacuum blotting (in which DNA/RNA fragments are transferred onto the filter *in vacuo*), and electroblotting (transfer in an electric field). These blotting methods are widely practiced for polyacrylamide gels whose small pore size does not permit effective capillary transfer. Both these methods require specified, commercially available equipment (e.g. Mighty Small Transphor Tank Transfer Unit from Hoefer, VacuGene XL Vacuum Blotting System from Pharmacia, or equivalents) and must be performed according to the manufacturer's instructions.

We recommend the use of *nylon filters* both in Southern and Northern transfer experiments. Compared to its alternative, nitrocellulose, nylon filters are much easier to handle because of their high tensile strength and inherent hydrophility. Nylon binds nucleic acids covalently and if stripped correctly, the filter can be reprobed several times. Using radiolabeled probes, both the uncharged nylon (Hybond-N, Amersham) and positively charged nylon filters (Hybond-N⁺, Amersham) yield good and reproducible results. The use of positively charged nylon filters is especially recommended for transfer in alkali conditions and for nonradioactive detection applications.

Procedure

All steps are carried out at room temperature.

1. After the electrophoresis is complete, stain the gel with ethidium bromide, examine and photograph it in ultraviolet light with the transparent ruler laid alongside the gel. Wear gloves, and keep all the surfaces the gel touches clean. Try not to illuminate the gel in UV light for more than a minute.

2. Place the gel onto a clean glass dish. Cut off the bottom left-hand corner of the gel. This serves as an orientation marker, ensuring that in subsequent hybridization procedures the DNA-carrying side of the filter is recognizable. Make a cut exactly along the line of slots and discard the upper part of the gel.

Depuration (Optional) An additional depuration step is required only if analyzing DNA fragments larger than 15 kb, prior to alkali denaturation. After step 2, add several volumes of 0.2 M HCl to the dish containing the gel, and soak for 15 min with agitation. Rinse with deionized water and proceed with denaturation.

Denaturation 3. Add 5 gel volumes of denaturation solution (0.5 M NaOH, 1.5 M NaCl) and soak for 30 min at room temperature with gentle agitation on the rotary platform.

Note. If using depuration, replace with fresh denaturation solution after 20 min and agitate for further 20 min.

Neutralization 4. Pour off the denaturation solution and rinse briefly with deionized water. Add 5 vol of neutralization buffer (0.5 M Tris-HCl 7.5, 1.5 M NaCl) and agitate as before for 20 min. Replace with fresh neutralization solution and agitate for further 20 min.

Note. We do not recommend the use of Tris-HCl with higher pH for this purpose.

5. Meanwhile, set up the transfer support (Fig. 6.1). Place a glass plate a bit wider than the gel over the support in the blotting reservoir. Put two pieces of Whatman 3MM paper over the glass plate so as to form a "bridge". Fill the reservoir with the transfer buffer and wet the bridge.

Transfer Setup

6. Cut a piece of nylon filter (either uncharged or positively charged, Hybond-N$^+$, Amersham or equivalent) of the same size as the gel. Handle the filter by the edges only, using blunt-ended forceps. Wear gloves. Wet the filter in deionized water, then submerge it in transfer buffer for 5 min. Cut off one corner of the filter.

7. Pour off the neutralization solution. Flood the bridge generously with transfer buffer (this assists in positioning of the gel) and place the gel onto the bridge. Squeeze out air bubbles from beneath the gel by rolling the glass pipet over the gel. Apply four strips of Parafilm along the edges of the gel. This ensures that paper towels will not touch the bridge and the liquid flow from the reservoir will go through the gel.

Position the Gel and Filter

8. Apply several ml of transfer buffer to the surface of the gel. Carefully position the wetted filter onto the gel so that the cut corners match and the edge of the filter touches the line of slots (this will later help to estimate the migration rate of hybridizing fragments). Ensure that the Parafilm strips remain on the edges of the gel. Try to place the filter precisely at once. Roll out any air bubbles from beneath the filter.

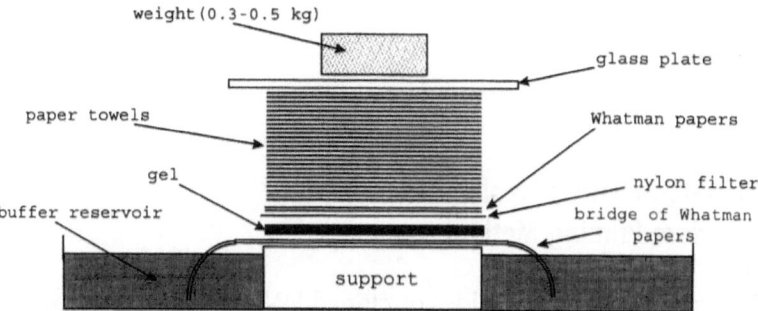

Fig. 6.1. A typical capillary blotting setup

9. Wet 2–3 pieces of 3MM paper of the same size as filter in transfer buffer and place them onto the filter. Cut paper towels to the same size and put these onto the 3MM papers to form a stack 7–8 cm high.

Transfer

10. Put a glass plate on the top and a weight (300–500 g) on it to hold the paper towels in good contact. Let the transfer proceed overnight.

Remove the Filter

11. Remove the paper towels (these should be wet) and 3MM papers above the filter. Using forceps, take off the filter and rinse it in 2x SSC to remove any adhering agarose. Place the filter onto a sheet of 3MM paper and allow to air dry for 10 min.

Note. Efficiency of UV cross-linking depends on the dryness of the filter. Always dry the filter in the same way.

Immobilize the DNA onto the Filter

12. Attach the DNA covalently to the nylon filter by irradiating the DNA-carrying side of the filter in a UV cross-linker (UV Stratalinker 1800 from Stratagene; follow the manufacturer's instructions).

Note. If using a laboratory UV transilluminator as UV light source, 1–3 min of exposure time will be adequate for most devices. However, for best results the optimal cross-linking time should be determined empirically for each particular transilluminator. This calibration should be done regularly, as the energy of UV lamps changes with their age.

13. The filter is ready for hybridization. For storage, place the filter between two sheets of 3MM paper and keep it at room temperature, protected from light.

6.3
Preparation of the Radiolabeled Probe by Random Priming Method

DNA probes (oligonucleotides or cloned DNA fragments) as well as RNA probes (transcribed in vitro from special vectors containing RNA polymerase promoters, see Materials in Chap. 3) can be used to probe complementary sequences on Southern and Northern blots. The detection methods can be either radioactive or nonradioactive (e.g. using digoxygenin, biotin etc.). Despite that the latter have the advantage of being safe, most consistent results are still obtained using radioactively labeled probes. For DNA labeling, nick translation (Rigby et al. 1977) and oligolabeling (Feinberg and Vogelstein 1983) are the most commonly used methods. The oligolabeling (or random priming) method utilizes *random hexanucleotides* which can prime DNA synthesis in vitro from any linear DNA template and yields probes with *highest specific activity* ($>10^9$ cpm/μg DNA). It can be used both for the detection of DNA as well as RNA sequences. To reduce the noise/signal ratio in hybridization, it is preferable to label only the specific sequence (e.g. the insert only, not the entire plasmid containing vector sequences also).

Caution. Work with radioactivity should be carried out in the designated areas, taking care to avoid any *radioactive contamination*. Monitor yourself and your working place frequently. *Follow strictly* the approved guidelines and regulations of the local radiation protection supervisor.

Procedure

1. Take 50–100 ng of the DNA fragment to be labeled and make the volume of the DNA up to 10 μl with deionized water. **Labeling the Probe**

Note. Using larger amounts of DNA does not result in higher radioactive activity of the product.

2. Denature DNA at 95 °C for 3–5 min and chill quickly on ice.

3. Add the following components on ice to the denatured DNA:
   5 μl OLB buffer
   1 μl dATP, dGTP, dTTP (2.5 mM of each)
   0.5 μl 20 mg/ml BSA
   50 μCi [α-^{32}P]dCTP
   2–5 U Klenow fragment of DNA polymerase I

Note 1. When fresh isotope is used, usually 50 μCi of [α-^{32}P]dNTP (3000 Ci/mmol, 10 μCi/μl) consists of 5 μl. If a different volume of the isotopically labeled nucleotide is used, correspondingly smaller or larger amounts of deionized water must be added to the DNA in step 1.

Note 2. It is recommended to use [α-^{32}P]dCTP due to the higher incorporation efficiencies, but other labeled deoxynucleotide triphosphates can be also used at will. Be sure to make necessary changes to the nonlabeled dNTP recipe.

4. Incubate the reaction at 37 °C for 60 min. If necessary, the reaction can proceed overnight at room temperature.

Purification of the Labeled Product

Usually the labeled probe can be used for hybridization without further removal of unincorporated label. If the purification of the labeled probe is required (due to the low incorporation levels or high background), it can be achieved by using Sephadex G-50 spun columns (Sambrook et al. 1989) or commercially available purification columns (NICK columns from Pharmacia Biotech).

Several oligolabeling kits now on the market (Oligolabeling Kit from Pharmacia Biotech, Prime-a-Gene Kit from Promega etc.) are convenient to use and ensure high incorporation efficiencies with extremely short incubation time. Follow the manufacturer's instructions.

6.4
Hybridization of the Radiolabeled Probe to the DNA Blot

A hybridization experiment is based on the principle that labeled nucleic acid molecules (probe) can form base-pairs with the *complementary nucleic acid* molecules immobilized on the filter. At first the filter is incubated in the hybridization solution without the probe (*prehybridization*). This step blocks *nonspecific binding* of the probe to the filter surface. Subsequently, the filter is incubated in fresh hybridization solution in presence of the labeled probe at conditions that permit the probe to bind to its homologous sequences (not only 100 % identical but even weakly homologous) on the filter. The nonspecific signal is eliminated gradually during the washing steps until the desired level of specificity of the probe is achieved. In this protocol we describe the experiment using a hybridization oven with a roller. Although hybridization in the heat-sealable plastic bags is even more commonly practiced, the hybridization technique in an oven is advantageous. It is completely free from the side-effects characteristic of the bag experiment, with the filters sticking to each other and to the bag wall. As little as 2 ml of hybridization solution can be used, in turn reducing the costs for purchase and disposal of radiochemicals. We find the use of the hybridization oven safe, convenient and economic for a lab routinely working with radioactive probes.

Procedure

1. Prewarm the hybridization oven and bottle to 65 °C. Prepare hybridization solution and if it contains particles, filter it through a cellulose acetate filter with pore size 0.45 µm (SFCA syringe filter from Nalgene).

2. Immerse the filter containing immobilized DNA in 6× SSC.

Note. Use gloves and blunt-ended forceps.

3. Shake off excess liquid and place the filter, DNA-side up, in the hybridization bottle and add about 0.5–1 ml of hybridization solution per 10 cm^2 of filter.   **Prehybridization**

Note. If using large filter or several filters, roll it (them) together with a piece of hybridization net (Biometra Hybridization Mesh). This will ensure even distribution of hybridization solution across the filter(s).

4. Incubate the bottle for 1–3 h in the hybridization oven with rotation at 65 °C (prehybridization).

5. Denature the radiolabeled DNA probe (specific activity $10^8–10^9$ cpm/µg, in amount of 1–2 µCi/ml hybridization solution) by heating it at 98–100 °C for 5 min. Chill on ice.   **Denature the Probe**

Note. This step should be done in conjunction with the next step. There is no need to denature single-stranded probes (e.g. RNA probes).

6. Pour off the prehybridization solution from the bottle and replace with prewarmed hybridization solution. Add the radiolabeled, freshly denatured probe to the bottle. Incubate with rotation overnight.   **Hybridization**

Note. If analyzing DNA with low complexity (e.g. plasmids, cosmids, PCR products), the hybridization solution need not be changed and the denatured probe can be added directly to prehybridization solution.

Wash the Filter

7. Pour the hybridization solution containing the probe into a container (this solution may be stored at −20 °C and reused) and quickly immerse the filter in a dish containing 2× SSC/0.5 % SDS. Incubate for 10 min at 25 °C with gentle agitation.

Note 1. The wash of filter can be done in the bottle, if the filter is small and its edges do not overlap. Otherwise it would be difficult to reduce the background evenly.

Note 2. Do not let the filter to dry out at any step of wash.

Note 3. Wash solutions should be treated as radioactive waste.

8. Replace the wash solution with 0.2× SSC/0.1 % SDS and incubate for 30 min at 50 °C with agitation. Change the wash solution after 15 min (wash at medium stringency conditions).

Note. Follow the efficiency of washes on the filter with hand-held monitor (Berthold LB 1202). There should be only a weak background radioactivity if any in the regions containing no DNA.

9. Replace with 0.1× SSC/0.1 % SDS and agitate at 65 °C for 30 min (wash at high stringency conditions).

Note. If the target DNA sequence is not identical to the probe (heterologous hybridization), do not perform high stringency wash. You may decrease the temperature of the final wash or use higher SSC concentration (e.g. 0.2× SSC) in the wash solution or combine them both. A useful rule of thumb – reduce the wash temperature by 1 °C per 1 % of mismatches (between the probe and the target sequence; Ausubel et al. 1987).

10. After the final wash, blot the filter briefly on a piece of Whatman 3MM paper to remove excess liquid. Place the damp filter between two sheets of Saran Wrap. Filter is now ready for autoradiography.

Note. Do not allow the filter to dry out before packing it into the Saran Wrap if you wish to strip and reprobe the filter.

Comments

Heat-Sealable Bags. If performing hybridization in heat-sealable plastic bags, ensure that the filter is correctly prewetted, covered entirely with the hybridization solution and can move freely in the bag. Use preheated hybridization solution, this reduces the risk of forming air bubbles within the bag. Do not hybridize more than two filters in the same bag. Even then lay the filters into the bag one by one, mounting them wholly in hybridization solution. Take care to minimize radioactive contamination while sealing the bag. Hybridization is best performed in a water bath with slow agitation.

Hybridization Analysis of Cloned DNA as the Target. The protocol presented above assumes that detection of a single copy sequence in a highly complex DNA (e.g. mammalian genomic DNA) is desired. However, in this ultimate case a 1 kb single copy target sequence makes only 3 pg in 10 µg of mammalian genomic DNA. Hybridization applications in analysis of cloned DNA (plasmids, cosmids etc.) do not require this degree of sensitvity, as the amount of target DNA can be even hundreds of nanograms. In this case the hybridization technique may be greatly simplified, by reducing the prehydridization time to 10 min and hybridization time to 2 h. The amount of probe used may be 0.05 µCi per ml of hybridization solution. In turn, the time for transfer of DNA from the gel to the filter may be shortened to 2 h.

6.5
Autoradiography

Autoradiography is a technique to visualize and quantitate radioactive molecules by exposure to X-ray film. Radioisotopes can be detected by direct autoradiography or by converting first the emitted energy into light and imposing light onto the X-ray film. The latter is usually achieved by use of screen(s). Direct autoradiography is less sensitive regardless of the isotope used. Energy emitted by isotopes with low energy ionizing radiation (^{3}H, ^{14}C, ^{35}S) is largely absorbed internally within the sample and does not reach the X-ray film. On the other hand, high energy radiation from ^{32}P or ^{125}I passes through the film so that only a small proportion of the energy is captured by the film. For nucleic acid hybridization ^{32}P is the radioactive isotope of choice in most cases, and we focus on autoradiography of radioactive molecules emitting highly energetic particles, i.e. ^{32}P.

Intensifying screens, used to detect radioactive isotopes indirectly, incorporate dense, inorganic scintillators. These scintillators efficiently absorb high energy emissions and emit photons in response to this absorption. The generated light produces a photographic image on X-ray film. For ^{32}P an intensfying screen increases the sensitivity of detection approximately tenfold. The disadvantages of intensifying screens are that they decrease the resolution of the obtained image as the secondary scintillations disperse, and that the film response to light is nonlinear. The latter problem can be overcome by pre-flashing the film using commercially available devices (Sensitize, Amersham). As photographic emulsion on X-ray film is disproportionally insensitive to very low intensities of light, autoradiography using the intensifying screen(s) should be carried out at −70 °C, as at this temperature the sensitivity of radiography film to low intensity of light is much greater (Koren et al. 1970).

Procedure

1. In the darkroom with red safe light, place the filter, packed in Saran wrap, in a cassette with intensifying screen(s) and overlay with suitable X-ray film (Hyperfilm-MP, Amersham or equivalent) so that the film remains between the screen and the filter. Close the cassette and place at −70 °C to expose.

Note. When using the cassette without intensifying screens, the autoradiography can be carried out at room temperature. The exposure time will be depend on the number of counts on the filter (as estimated by radioactivity minimonitor) and can range from 1 h to a couple of weeks.

2. To develop the film, remove the cassette from the −70 °C freezer and either develop the film immediately or allow to thaw completely before opening in the dark room. Under the red safe light, open the cassette, clip the film onto a suitable hanging frame and place in the developer tank for 4–8 min.

3. Wash briefly in the water tank.

4. Transfer the film to the fix tank and leave for 5–10 min.

5. Wash the film first under running tap water and then rinse with deionized water.

6. Air dry the film.

Note. Several companies are offering ready-made developing and fixation solutions for X-ray films which can be used instead of self-made solutions. Furthermore, automatic developers are available on the market (e.g. Kodak X-Omat M43A) where developing, fixing, washing and drying of the film is carried out automatically.

A complete alternative to autoradiography is the use of phosphoimagers (e.g. PhosphoImager SI, Molecular Dynamics), where the filter is exposed to a specific screen and the computer detects the emission signal of radioisotopes.

6.6
Electrophoresis and Transfer of RNA

Working with RNA demands some extra precautions, as RNA is rapidly degraded in the presence of RNases. As ribonucleases are extremely stable (they can remain active even after exposure to a temperature of over 100 °C), the possibility of contaminating the working solutions or samples should be kept minimal. First of all, the most probable source for the RNase contamination is the researcher himself. Therefore we recommend wearing gloves and changing them frequently when handling any reagents or vessels used to work with RNA. Next, common DNA isolation procedures involve use of RNases, leading to contamination with it everywhere: tubes, racks, solutions, electrophoresis tanks, and gel trays. It is desirable to put aside a set of equipment, glass- and plasticware, and the necessary autoclaved solutions, and use them exclusively for RNA work. Inactivation of RNases minimizes the risk of introducing RNases from exogenous sources. Glassware should be baked at 200 °C for 6 h. Equipment not suitable for autoclaving can be treated with 0.5 M NaOH prior to use. An efficient inactivator of RNases in solutions is diethyl pyrocarbonate (DEPC). For DEPC treatment of deionized water and other working solutions, add DEPC at a final concentration of 0.1 %, shake well and incubate overnight at room temperature in a fume hood. Autoclave for a least 20 min to remove DEPC (otherwise DEPC will inactivate RNA).

Precautions for Work with RNA

Note. Do note treat solutions containing Tris with DEPC.

Caution. DEPC is a suspected carcinogen. A fume hood must be used.

■ Procedure

Electrophoresis of RNA Through Denaturing Gels

Pouring of the Denaturing 1% Agarose Gel

Note. The following protocol is for pouring 50 ml RNA gels. If a particular gel casting system requires gels with a different volume, change all amounts proportionally.

1. Weigh 0.5 g agarose (SeaKem LE from FMC BioProducts or equivalent) into a 100 ml flask and add 5 ml of 10× MOPS buffer and 37 ml of double-distilled water.

Note. For separation of small RNA fragments (≤ 0.8 kb), higher agarose concentrations (up to 2%) and special agaroses (NuSieve 3:1 from FMC BioProducts or equivalent) can be used.

2. Dissolve agarose in microwave oven.

Note. Bring to boil and then set the power of the oven to the lowest level or boil with short intervals of some seconds each, gently swirling the mixture after each interval. Otherwise the agarose solution will boil over.

3. Cool the agarose to 55–65 °C (until you can hold the flask with the unprotected hand). Add 8 ml of concentrated formaldehyde, mix well by swirling and pour the gel using a suitable mini-gel former immediately.

Note 1. To avoid contamination of the gel-forming unit with RNases, it is recommended to treat the unit with 0.5 M NaOH.

Note 2. Concentrated formaldehyde is usually sold as a 37–40% v/v solution (37% formaldehyde solution from Merck). The concentrated formaldehyde should be handled and the corresponding agarose/formaldehyde gel should be poured in a fume hood.

Denaturation of RNA Samples

4. Mix your RNA samples with 16.5 µl of RNA loading buffer. The volume of the sample should be between 1–8 µl.

Note. For successful detection of low abundance transcripts from the total RNA sample, approx. 10–30 µg of total RNA should be taken for each lane. When the poly(A)$^+$ fraction of the total RNA is analyzed, approx. 1 µg of the RNA sample should be enough. For other purposes, the amount of the RNA needed could be considerably smaller.

5. Denature the RNA samples at 65 °C for 2–3 min and chill on ice immediately.

6. Load the samples on the gel.

Note. If the size of the separated RNA molecules has to be esti-mated, suitable RNA molecular weight markers must be loaded onto the gel together with other samples. Either commercially available RNA size markers (Promega RNA markers or equival-ent) or an RNA sample with rRNA bands of known size can be used.

7. Run the gel in 1× MOPS buffer at up to 4–5 V/cm (measured as distance between the electrodes) until the dye reaches the bottom fourth of the gel.

Note. Treat the gel running equipment prior to use with 0.5 M NaOH to destroy RNases.

8. Photograph the gel on the UV transilluminator. If the blotted RNA loading dye does not contain ethidium bromide, these lanes should be cut off from the gel before photographing.

Transfer of RNA to a Nylon Membrane, Hybridization, and Autoradiography

Northern blotting or transfer of RNA to a membrane is per-formed according to the ordinary capillary flow method (Sect. 6.2). As RNA is electrophoresed in the denaturing condi-tions, the gel does not need any denaturation and neutralization steps. However, it is advised to remove the formaldehyde before blotting. We recommend the use of nylon filters (uncharged or positively charged) for RNA transfer.

1. After electrophoresis and photography, soak the formalde-hyde/agarose gel for 20 min in 10 vol of deionized water.

Note. Do not perform depurination, denaturation and neutral-ization procedures.

2. At the same time, set up the transfer support for blotting RNA onto the nylon filter, as described for DNA (Sect. 6.2, steps 5–10).

Note. It is recommended to treat the blotting equipment with 0.5 M NaOH before use to avoid contamination with RNases.

3. Carry out the transfer, dismantle the blot and dry the filter as described for the Southern transfer.

Cross-Linking of RNA to Nylon Membrane

4. Cross-link RNA to the nylon membrane. If crosslinking is performed on the UV transilluminator, a considerably shorter exposure to UV light is recommended when compared to the crosslinking of DNA fragments. We have commonly used a 30-s exposure time for the usual laboratory UV transilluminators. When Stratalinker is used, the same auto cross-link program as for DNA can be used.

5. Carry out prehybridization, hybridization, filter washing and autoradiography as described for DNA.

Results

● DNA digestion and electrophoresis
Digestion of genomic DNA with enzymes that recognize hexa-nucleotide sequences produces a smear of DNA fragments on the gel ranging from 1 to 20 kb in length. Enzymes which recognize sequences containing CpG dinucleotides (*Nru*I, *Bss*HII, *Eag*I, *Sac*II etc.) rarely cleave mammalian DNA and are not advised for performing conventional Southern blot analysis. The CpG-containing-enzymes are specifically used in pulse-field gel electrophoresis applications.

DNA fragments are usually successfully separated under the described conditions when the proper agarose concentration is used. The bands of DNA molecular weight markers must be sharp and well resolved. The intensity of smaller DNA fragments may be lower in gels containing ethidium bromide, as during electrophoresis ethidium bromide migrates towards the cathode and runs out from the bottom part of the gel.

● DNA blotting
After dismantling the blot, the gel can be restained with ethidium bromide to assess the efficiency of transfer. Another way to follow the success in transferring DNA to the filter is to examine whether the filter contains spots from the transfer of the tracking dyes.

● Autoradiography of DNA filters
Analysis of cloned DNA by Southern blotting usually needs an exposure of 1–2 h at −70 °C. Using probes of 0.2–1 kb with

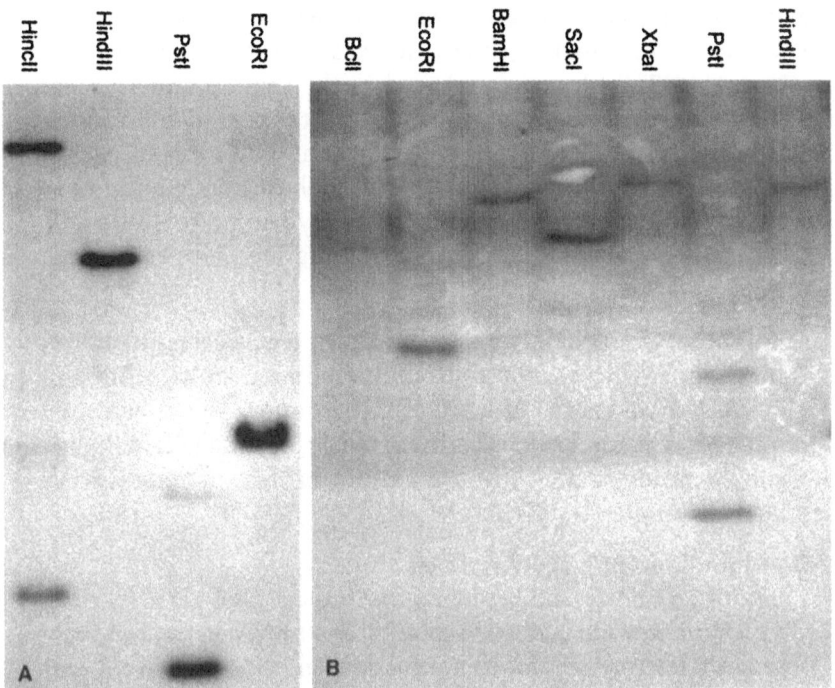

Fig. 6.2.A–B. Southern blot analysis of mouse genomic DNA. DNA was cut with restriction endonucleases as indicated *above* the lanes. **A** A successful Southern blot experiment. **B** False result caused by contamination with DNA able to hybridize to the probe

high specific activity that are identical to the target sequence, single-copy sequences in genomic DNA should be detected by 1–2 days of exposure at −70 °C. With shorter probes or when detection of less homologous sequences is required, the exposure time may be up to 2–3 weeks.

Figure 6.2 shows results of a typical genomic Southern blot experiment. Mouse genomic DNA (15 µg) was digested with the restriction enzymes indicated above the lanes on the figure and electrophoresed through 0.8 % agarose/TBE gel. Filters were hybridized to an intronic fragment from the single copy gene encoding mouse ribosomal protein S6, labeled by random priming. Panel A shows the results of a successful Southern blot experiment. Characteristic of a single copy sequence, only one band hybridizing to the probe is detected in lanes where genomic DNA has been cut with *Eco*RI and *Hind*III. Two bands are visible in lanes of genomic DNA digested with *Hinc*II and *Pst*I. These enzymes have a recognition site within the probe sequence, thus the probe is divided between two fragments of genomic DNA. Panel B shows the false results of the similar experiment. The blot was prepared in the same manner except that the gel was electrophoresed in the tank used for routine plasmid analysis. Note the high background in the upper part of the filter. This indicates that the electrophoresis apparatus has been contaminated with DNA able to hybridize to the probe (likely positive plasmid DNA) that has migrated into the gel.

- RNA electrophoresis
  Poly(A)$^+$ RNA is barely visible on agarose gels even after careful staining with ethidium bromide (followed by destaining in deionized water to reduce the background). However, when total RNA is analyzed, clearly visible ribosomal RNA bands should be detectable when ethidium bromide stained gel is analyzed under UV light.

- Autoradiography of RNA filters
  Figure 6.3 shows the result of the typical Northern blotting experiment, where 10 μg of total tobacco RNA, extracted from various transgenic clones are loaded on each lane. RNA was electrophoresed in 1 % agarose/formaldehyde gel. Oligolabeled potato virus X cDNA has been used as a [^{32}P]-labeled radioactive probe. The film was exposed to the filter for 4 days at −70 °C using intensifying screens. One band is seen on every lane, corresponding to the potato virus X coat protein

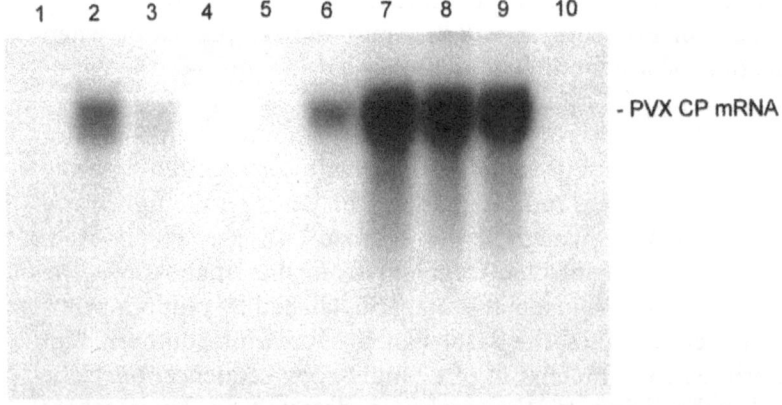

Fig. 6.3. Northern blot analysis of total RNA extracted from transgenic tobacco plants transformed with the cDNA encoding the coat protein of potato virus X. *Lane 1* RNA from nontransgenic control tobacco; *lanes 2–10* RNA from independent transgenic tobacco clones

transgene mRNA in transgenic tobacco plants. Note that the intensities of bands on different lanes vary, indicating that the expression levels of the transgene are different in different plant clones. Some samples have not hybridized to the potato virus X coat protein probe at all, indicating that these lines do not express detectable amounts of coat protein mRNA.

Troubleshooting

- DNA bands poorly resolved
  Prepare gel solution with optimal agarose percentage (see Table 6.1). Check that the buffer in the electrophoresis tank has not been exhausted during the run. The tank must be set horizontally and the buffer should cover the gel evenly. Ensure that no leakage has occurred during loading of the samples into the slots. If resolving large DNA fragments (more than 5 kb), pour a larger gel and run it for a longer time with low applied voltage (1–2 V/cm). Smearing of DNA band is commonly caused by overloading the sample (thus do not load more than 200 ng per band).

- Poor signal on film
  - Inefficient probe labeling. Purify the labeled product and count its radioactivity using a scintillation counter. If insufficient activity is determined (less than 10 % of label incorporated into probe), check the quality of the labeled DNA template and your labeling components.
  - Incomplete denaturation of the double-stranded probe. Denature the probe at least 3 min at 95 °C and chill on ice immediately.
  - Filter was washed at too high stringency conditions. Check the radioactivity of the filter after each washing step and go on with the autoradiography at the step when the nonspecific signal is washed away.
  - Inefficient transfer. Check that the Parafilm pieces were laid on the edges of the gel correctly preventing the blotting buffer being absorbed directly into the paper towels above. Nucleic acid transfer can be checked by restaining the gel with ethidium bromide.
  - Poor cross-linking. Check that UV transilluminator is correctly calibrated. Be sure that you have exposed the filter nucleic acid side down on the transilluminator.

- Check that the hybridization solution is correctly prepared.
- Try different hybridization and washing conditions.
  After prehybridization, replace with hybridization solution containing 6× SSC, 0.5 % SDS, 0.1 mg/ml salmon sperm DNA (Denhardt solution omitted) and labeled probe.
  Deionized formamide can be added to the hybridization mix. The prehybridization and hybridization should be carried out then at 42–65 °C, depending on the particular system.
- When RNA is analyzed, RNA degradation is the most likely cause of problems. Analyze your samples with the loading buffer containing ethidium bromide. If total RNA is used for the Northern analysis, in ethidium bromide stained samples two bands corresponding to the two major ribosomal RNAs should be clearly visible. However, for poly(A)$^+$ RNA the sample is barely visible after ethidium bromide staining. To avoid RNA degradation, use fresh buffers, alkali treated electrophoresis and blotting equipment and wear gloves throughout the experiment. Use separate solutions, tips, microcentrifuge tubes, and equipment only for the work with RNA. Treat buffers and solutions with 0.1 % diethyl pyrocarbonate.

Note. Do not treat solutions containing Tris with diethyl pyrocarbonate.
The best results will be obtained when the RNA is analyzed promptly after its isolation. When RNA storage is needed, it should be done at −70 °C or alternatively at −20 °C in pure formamide.

- High background signal on film
  - Carry out the high stringency washing of the filter, if you have not done it.
  - Inefficient prehybridization step. Use high quality blocking components to ensure the complete blocking of the free filter surface. Prepare freshly sonicated and denatured heterologous high molecular weight DNA.
  - Unincorporated label is giving the high background. Purify the labeled product and calculate the specific activity of the probe.
  - Particulate matter in the hybridization buffer. Heat hybridization buffer at 50–60 °C to dissolve possible precipitates and filter the buffer through 0.45 μm cellulose acetate filter (Nalgene SFCA).

- Residual agarose is adhering to the filter. Rinse the filter carefully with 2× SSC after the transfer.
- Part of or the whole filter has dried out during the hybridization or washing steps.
- Filters have stuck together during the hybridization or washing.
- Too small volumes of the washing solutions used.
- Try different hybridization and washing conditions.

  After prehybridization, replace with hybridization solution containing 6× SSC, 0.5% SDS, 0.1 mg/ml salmon sperm DNA (Denhardt solution omitted) and labeled probe.

  Deionized formamide can be added to the hybridization mix. The prehybridization and hybridization should be carried out then at 42–65 °C, depending on the particular system.

  We have occasionally used hybridization in a buffer containing dextran sulfate (1 M NaCl, 1% SDS, 0.1 mg/ml sonicated heterologous DNA, 10% dextran sulfate). Dextran sulfate is omitted from the prehybridization buffer. Washing of filters is carried out as described above. Dextran sulfate is a rate enhancer for probes larger than 200 base pairs. It should not be used with oligonucleotide probes.

- Extra bands
  - Too low stringency washing conditions. Keep washing with low salt washing buffer at high temperatures (up to 70 °C) until the radioactivity of the filter is acceptable (according to the minimonitor).
  - Incomplete digestion of target DNA. Refer to troubleshooting guide above.
  - The probe contains nonspecific sequences. Purify the fragment that contains only the desired sequence.
  - When analyzing RNA, we have observed that if nonspecific hybridization of some RNA species in Northern blotting becomes a problem, alkaline transfer in the presence of 10 mM NaOH (Löw and Rausch 1994) may overcome the problem.

References

Ausubel FM, Brent R, Kingston RE, Moore DD, Smith JA, Seidman JG, Struhl K (1987) Current Protocols in Molecular Biology. John Wiley & Sons, New York

Bailey JM, Davidson N (1976) Methylmercury as a reversible denaturing agent for agarose gel electrophoresis. Anal Biochem 70:75–85

Feinberg AP, Vogelstein B (1983) A technique for radiolabeling DNA restriction endonuclease fragments to high specific activity. Anal Biochem 132:6–13

Koren JF, Melo TB, Prydz S (1970) Enhanced photoemulsion sensitivity at low temperatures used in radiochromatography. J Chromatogr 46:129–131

Lehrach H, Diamond D, Wozney JM, Boedtker H (1977) RNA molecular weight determinations by gel electrophoresis under denaturing conditions, a critical reexamination. Biochemistry 16:4743–4751

Löw R, Rausch T (1994) Sensitive, nonradioactive Northern blots using alkaline transfer of total RNA and PCR-amplified biotinylated probes. BioTechniques 17:1026–1030

McMaster GK, Carmichael GG (1977) Analysis of single- and double-stranded nucleic acids on polyacrylamide and agarose gels by using glyoxal and acridine orange. Proc Natl Acad Sci USA 74:4835–4838

Rigby PWJ, Dieckmann M, Rhodes C, Berg P (1977) Labeling deoxyribonucleic acid to high specific activity in vitro by nick translation with DNA polymerase I. J Mol Biol 113:237–251

Sambrook J, Fritsch EF, Maniatis T (1989) Molecular Cloning: A Laboratory Manual. CSHL Press, New York

Southern EM (1975) Detection of specific sequences among DNA fragments separated by gel electrophoresis. J Mol Biol 98: 503–517

Suppliers

American Can Company
Greenwich, Connecticut 06830
USA

Amersham International plc
Little Chalfont
Buckinghamshire HP7 9NA
ENGLAND
http://www.amersham.co.uk

EG&G Berthold
Calmbacherstrasse 22
D-75323 Bad Wildbad
GERMANY

Biometra Ltd
Whatman House
St Leonard's Road
20/20 Maidstone
Kent ME16 0LS
ENGLAND
http://biometra.gbnet.co.uk

Fermentas AB
V. Graiciuno 8
LT-2028 Vilnius
LITHUANIA
http://www.fermentas.com

FMC BioProducts
191 Thomaston Street
Rockland, Maine 04841
USA
http://www.bioproducts.com

Hoefer Pharmacia Biotech Inc.
654 Minnesota Street
San Francisco, California 94107–0387
USA
http://www.hpb.com

Hybaid Ltd
111–113 Waldegrave Road
Teddington
Middlesex TW11 8LL
ENGLAND
http://www.hybaid.co.uk

E. Merck
Frankfurter Strasse 250
D-6100 Darmstadt
GERMANY
http://www.merck.de

Molecular Dynamics
928 East Arques Avenue
Sunnyvale, California 94086-4520
USA
http://www.mdyn.com

Nalge Co.
75 Panorama Creek Dr.
P.O. Box 20365
Rochester, New York 14602-0365
USA
http://nalgenunc.com

Polaroid Corp.
Cambridge, Massachusetts 02139
http://www.polaroid.com

Pharmacia Biotech AB
S-751 82 Uppsala
SWEDEN
http://www.biotech.pharmacia.se

Promega Corporation
2800 Woods Hollow Road
Madison, Wisconsin 53711-5399
USA
http://www.promega.com

Stratagene
11011 North Torrey Pines Road
La Jolla, California 92037
USA
http://www.stratagene.com

Chapter 7

Immunological Methods for Analysis of Recombinant Proteins

IRINA SOMINSKAYA[*][1] AND KASPARS TARS

Introduction

We want to introduce researchers to techniques that help to solve some problems in the work of the molecular biologist. After transformation of recombinant DNA in *E. coli* cells many clones are usually produced, and the same situation appears if recombinant DNA expression libraries are available. Furthermore, if appropriate monoclonal or polyclonal antibodies are available, the method of immunoscreening of colonies for direct immunological detection of translational products of cloned genes can be used. Selected clones could be chosen for further studies such as determination of their primary structure by DNA sequencing and for characterization of an appropriate expressed protein.

Principles and Application

Fractionation of proteins on polyacrylamide gels is one of the primary means of their characterization because it is simple and rapid. The most widely used technique is SDS-polyacrylamide denaturing gel electrophoresis (SDS-PAGE; Laemmli 1970). One-dimensional gel electrophoresis under denaturing conditions separates proteins based on molecular size as they move through a polyacrylamide gel matrix towards the anode. As a general rule one uses 5 % gels for proteins of molecular mass of 60–200 kDa, 10 % gels for proteins of 16–70 kDa, and 15 % gels for proteins of 12–45 kDa. Typically, the separating gel size is in the range 14×14 cm. Smaller formats (7×10 cm) are also popular, but do not offer the same level of resolution when compared to larger-format gels, but minigels use less reagent and separate proteins much faster. For 0.75 mm×14 cm×14 cm gels with

[*] Corresponding author: Irina Sominskaya, phone: +371–2–426126; fax: +371–2–427521; e-mail: irina@biomed.lu.lv
[1] University of Latvia, Biomedical Research and Study Centre, Ratsupites Str. 1, LV1067 Riga, Latvia

0.8 cm wide wells separation of a complex mixture of 25 to 50 µg of total protein is recommended when staining is done with Coomassie blue, but only 1 to 10 µg of total protein is needed for samples containing one or only a few proteins. If silver staining is used, 10- to 100-fold less protein mixture can be applied. For minigels (0.75 mm×7 cm×10 cm) Coomassie blue staining of a complex mixture needs 20 to 25 µg of protein, or 1 to 5 µg of highly purified proteins or 10- to 100-fold less for silver staining. The gel consists of two parts: the upper gel which is called the stacking gel and the lower gel which is called the resolving gel. The stacking gel is a large pore size gel, the buffer (1.25 M Tris-HCl, pH 6.8) in which this gel is made contains an anion whose electrophoretic mobility is greater than that of protein, while the electrode buffer contains an ion whose mobility is less than that of protein. During electrophoresis the "leading ion" (chloride) in the stacking gel moves faster than the protein and leaves behind it a zone of lower conductivity. The higher voltage gradient of this zone causes the protein to move faster and to "stack" at the boundary between the leading and terminating ions (glycine). The second gel layer (resolving gel) has a smaller pore size and it is prepared in a buffer of higher concentration and pH (1.875 M Tris-HCl, pH 8.8). The mobility of the terminating ion increases and its boundary moves ahead of the protein. Protein is resolved into individual bands according to size.

If you have a complex mixture of proteins and want to detect only one of them you should use one of the immunodetection protocols (immunoblotting). This method uses polyclonal or monoclonal antibodies and is based on an antigen-antibody interaction. This method includes electrophoretic transfer of proteins from the polyacrylamide gel to a nitrocellulose membrane, subsequent saturation of free spaces on the membrane with neutral protein solution (for example BSA solution, dry skimmed milk solution etc.) and incubation of the membrane with either monoclonal or polyclonal antibody, as appropriate. By use of the Western blotting technique as little as 1 ng of denatured form of protein antigen recognized by its antibody is detectable.

Materials

- Nitrocellulose membranes (BIO-RAD, cat. no. 162–0148)
- Polyethylene sheets
- Cellophane membranes

- Vertical electrophoresis unit (LKB, cat. no. 2050–001)   **Equipment**
- Electrophoretic transfer unit (LKB, cat. no. 2051–001)
- Power supply (LKB, cat. no. 2121–001)
- Gel Slab Dryer (Hoefer, cat. no. C980080)
- Microcentrifuge (Eppendorf, cat. no. M7282)

- Lysis buffer   **Buffers**
  60 mM Tris-HCl, pH 6.9
  2 % β-mercaptoethanol
  2 % SDS
  0.05 % Bromphenol blue
- Acrylamide stock solution
  29.1 % acrylamide
  0.9 % bis-acrylamide
- 10× electrophoresis buffer
  0.25 M Tris-HCl, pH 8.8
  1.92 M glycine
  1 % SDS.
- Resolving gel
  50 % acrylamide stock solution
  0.375 M Tris-HCl, pH 8.8
  0.1 % SDS
  Before pouring add 0.01 vol 10 % ammonium persulphate and
  0.001 vol TEMED.
- Stacking gel
  16 % acrylamide stock solution
  0.125 M Tris-HCl, pH 6.8
  0.1 % SDS
  Before pouring add 0.004 vol 10 % ammonium persulphate
  and 0.001 vol TEMED.
- Silver staining reagent
  0.0012 % NaOH
  0.3 % NH_4OH
  0.2 % $AgNO_3$
- Developing reagent
  0.005 % citric acid
  0.037 % formaldehyde
- 10× transfer buffer
  0.25 M Tris-HCl, pH 8.3
  1.92 M glycine
- 1× transfer buffer
  Dilute 10× buffer ten times, add methanol to final concentration of 20 %.

- Ponceau S solution
  0.5 % Ponceau S
  1 % acetic acid
- 0.15 M phosphate buffer
  0.10 M Na_2HPO_4
  0.05 M KH_2PO_4
- PBS
  0.15 M NaCl
  0.15 M phosphate buffer
- Tris-HCl/NaCl solution
  10 mM Tris-HCl, pH 7.5
  150 mM NaCl
- Staining solution
  o-dianizidine (10 µg/ml)
  5 % ethanol
  10 mM Tris-HCl, pH 7.5
  150 mM NaCl
  0.03 % H_2O_2

Procedure

Gel Electrophoresis of Proteins

Sample Preparation

1. Centrifuge at 10 000 rpm for 3 min 1.5 ml of *E. coli* cells grown overnight in 4 ml of LB medium containing ampicillin (50 µg/ml) Remove supernatant completely.

2. Resuspend cell pellets by vortexing and immediately add 200 µl of lysis buffer for each five optical units at 540 nm.

3. Vortex vigorously and place samples into boiling water for 7 min.

4. Vortex once more and allow to cool to room temperature.

5. Centrifuge samples for 2 min at top speed before loading protein solution onto the gel.

Note. Lysis procedure is the same for other sources of protein (column fractions, etc.) which you can use instead of cell pellets.

Preparation of a Gel

Note. Acrylamide is a neurotoxin. Wear gloves during this procedure to avoid skin contact.

1. Assemble glass-plate sandwich using two clean glass plates (11×14 cm) and two 0.75-mm spacers. Lock sandwich to the casting stand.

2. Prepare 10 ml of resolving gel (without TEMED) as described, add TEMED, gently mix and immediately pour the gel between glass-plates to a height of 11 cm for 14-cm glass plates.

3. Overlay the acrylamide gel with water-saturated n-butanol (height of layer to be about 1 cm).

4. After polymerization is complete, remove n-butanol, prepare and add 5 ml of stacking gel solution and insert the comb.

5. After polymerization carefully remove the comb, fill wells with buffer.

6. Using syringe load 5 µl of samples on gel.

7. Attach gel sandwich to upper buffer chamber. Fill lower buffer chamber with electrophoresis buffer and place sandwich into lower chamber. Fill upper chamber with electrophoresis buffer.

Electrophoresis

Electrophorese samples at 30 mA until bromophenol blue reaches the bottom of the gel (for 11 cm gels it takes 2–3 h). Electrophoresis can be carried out at room temperature.

After electrophoresis gel can be either stained (see below) or transferred to nitrocellulose membrane in order to perform Western blot procedure.

Silver Staining of the Gel

Note. Reagents for staining should be prepared prior to use. The silver reagent must be prepared in the sequence indicated in order to avoid the formation of precipitate. Use magnetic stirrer for preparing this reagent and add silver nitrate dropwise.

1. Wash gel with vigorous shaking with 100 ml of 50 % ethanol for 1 h.

2. Transfer gel to a tray containing silver reagent and stay for 15 min. Wear gloves during this procedure.

3. Wash gel four times for 15 min with distilled water.

4. Place gel into a second tray containing developing reagent.

5. Rinse gel with 10 % (v/v) acetic acid (to stop development).

6. Wash gel three times for 30 min with distilled water.

Staining with Coomassie Blue

1. Put the gel into solution of 0.125 % Coomassie Blue R250 containing 50 % methanol and 10 % acetic acid for 15 min with continuous shaking.

2. Destain the gel in 50 % methanol and 10 % acetic acid solution until background disappears.

3. Wash the gel with distilled water.

Drying of the Gel

1. Lay a wet cellophane membrane on the support sheet. Carefully lay the pretreated gel on top of the membrane. Gently smooth out any air bubbles between the membrane and the gel.

2. Lay the other membrane sheet and again smooth out any air bubbles. If more than one gel is to be dried, repeat these operations for each gel.

3. Place cover sheet over the gel, cover the whole assembly carefully with a heating cloth. Apply vacuum and gently press down the edges of the heating cloth so that it seals evenly by suction to the base plate.

4. Connect Gel Slab Dryer to the power supply. Turn it on and apply the appropriate voltage (24 V) for 1.5 h.

Western Blotting of Electrophoretically Separated Proteins

Transfer Procedure

1. Fill buffer tank with transfer buffer. Fill a tray big enough to hold the cassette with transfer buffer.

Note. Transfer buffer generally may be used several (2–3) times. However, it should not be stored in the unit. If you are not doing another run immediately, pour out the buffer and rinse the chamber with distilled water without removing the electrode panels.

2. Open the cassette. Place one half in the buffer tray and put foam sponge, filter paper, nitrocellulose membrane sheet, polyacrylamide gel, another filter paper and another sponge on it sequentially. Avoid trapping air bubbles between the transfer membrane and the gel.

3. Press gently to force out trapped air and then place the second half of the cassette on the top of the stack.

4. Press the two halves together gently. The cassette with its packing should hold the gel in firm contact with the transfer membrane without squashing the gel.

5. Lift the cassette sandwich and insert it into one of the sets of matching vertical slots in the end plates of the blot tank. The cassette should be loaded under buffer to avoid trapping air bubbles between the different layers. The gel in each cassette should be on the same side of the membrane so that all proteins from the gel will migrate from the gel toward the membrane when electric field is applied.

6. Place a magnetic stir bar into the buffer chamber and place the chamber on a magnetic stirrer. Set stirrer motor at a moderately low speed.

7. Connect the unit with power supply and turn on the power switch. Set the voltage to give a current reading of 200 mA.

8. When the transfer is complete (after about 1 h), turn off the power and remove the lid from the unit.

9. Lift out the cassette and open it carefully to remove the gel and transfer membrane. Discard the blotting paper. The foam sponge may be reused indefinitely.

10. Place the nitrocellulose membrane in Ponceau S solution for 5 min to stain proteins. Destain in water for 2 min. All proteins must appear as individual bands on membrane. Completely destain the filter by soaking it in water for about 10 min.

Note. This procedure is only to check quality of transfer and is not needed for further reactions.

1. Incubate the nitrocellulose membrane in PBS containing 1 % BSA for 1 h at room temperature in order to block free places on the membrane; agitate continuously.

Washing and Staining of Nitrocellulose Membranes

Note. All following procedures should also be done at room temperature.

2. Transfer membranes into plastic heat-sealable bags and add 1.0 ml of antibodies diluted with PBS containing 1 % BSA. Incubate overnight.

Note. Dilution rate of antibodies and conjugates depends on their concentration and nature and is determined experimentally (usually 1:100 to 1:10 000).

3. Wash membranes twice for 5 min each with PBS, continuously agitating.

4. Transfer membranes into new plastic bags and add 1.0 ml of horseradish peroxidase conjugate diluted with PBS containing 1 % BSA. Incubate for 2 h.

5. Wash membranes 3 times for 10 min each with PBS containing 0.1 % Tween 20 (FERAK, cat. no. 51601) and 10 min in Tris-HCl/NaCl solution, with gentle agitation.

6. Prepare staining solution. Transfer membranes into the staining solution and agitate continuously. When staining is complete (usually 2–5 min), rinse membranes with distilled water and dry. Coloured spots (in case of immunological screening) or bands (Western blot) on the membrane correspond to the proteins which reacted with the specific antibody.

Immunological Screening of Colonies

1. Pick the recombinant *E. coli* colonies to be screened onto master LB-agar plates using sterile toothpicks. Mark the plates asymmetrically. Put negative and positive control clones on each plate. Incubate the plates overnight at 37 °C.

2. Transfer the colonies from the master plates to nitrocellulose filters. Label the filters and mark them with a ballpoint pen.

3. Put the filters on agar plates and incubate at 37 °C. Bacterial colonies are grown on nitrocellulose filters until they reach a diameter of at least 1–2 mm. Letting the colonies grow larger than 2 mm does not appear to affect the assay. Store the master plates inverted and sealed in Parafilm at 4 °C.

4. Remove the nitrocellulose filters from the medium plates with forceps and soak the filters in chloroform for 3–4 min. Remove them from the chloroform and place on filter paper. Then store them for 2 min at room temperature and then 2 min at 37 °C in order to remove traces of chloroform.

5. Now do the same as described in Western Blotting experiment beginning from step 2 (see p. 141)

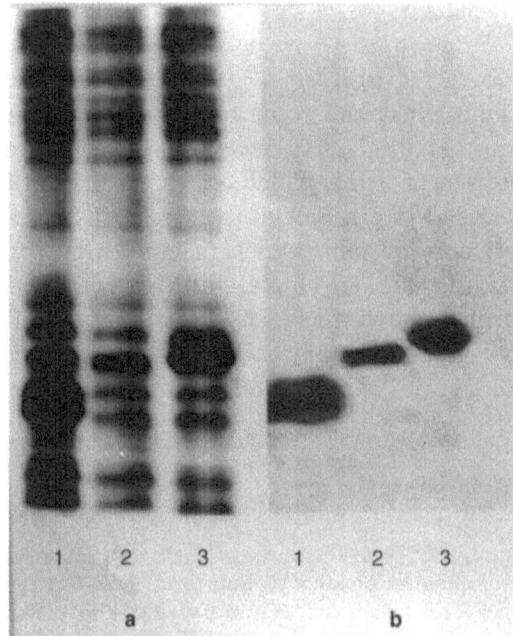

Fig. 7.1.a Electrophoresis of lysed *E. coli* cells with expression of recombinant phage *fr* native coat protein (*lane 1*) and same protein with foreign amino acid insertions (from hepatitis B antigene preS1 region; *lanes 2* and *3*). **b** Western blot with same proteins as in **a.** Only proteins which react with anti-*fr*CP antibodies are coloured

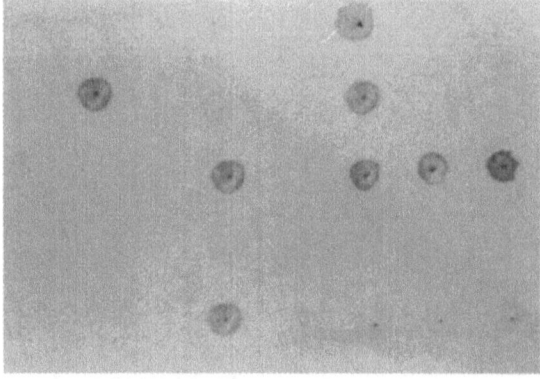

Fig. 7.2. Immunological screening experiment. Colonies expressing *fr* CP antigen appear as coloured dots on nitrocellulose membrane

Results

All methods were tested with bacteriophage fr coat protein
(Puskho et.al.) and some of its insertion mutants, expressed in
E. coli. Figure 7.1a shows electrophoresis of lysed cells with
expressed proteins, which appear as thicker bands than cellular
proteins. Western blot of the same lysates is shown in Fig. 7.1b,
where only proteins which react with anti-fr antibodies can be
seen. Immunological screening of colonies containing fr coat
protein is shown in Fig. 7.2. For immunological reactions polyclonal anti-rabbit antibodies against denatured fr phage were
used.

References

Laemmli UK (1970) Cleavage of structural proteins during assembly of the
    head of bacteriophage T4. Nature 227:680–685
Pushko P, Kozlovskaya T, Sominskaya I, Brede A, Pumpens P, Stankevica E,
    Ose V, Grens E (1993) Analysis of RNA phage *fr* coat protein assembly by
    insertion, deletion and substitution mutagenesis. Protein Eng 6:883–891
Towbin H, Staehelin T, Gordon J (1979) Electrophoretic transfer of proteins
    from polyacrylamide gels to nitrocellulose sheets: procedure and some
    applications. Proc Natl Acad Sci USA 76:4350–4354

Chapter 8

Protein Synthesis in Cell-Free Systems

Tiina Tamm[1] and Erkki Truve[*1]

Introduction

Cell-free extracts are widely used to identify and study the proteins encoded by cloned cDNAs, viral genomes or other exogenous mRNA, etc., as well as to investigate transcriptional and
translational control. In vitro translation is in many instances
the method of choice for the rapid analysis of uncharacterized
proteins, provided that the cDNA encoding for the protein is
available. It has been demonstrated that the in vitro translated
protein can be used for immunoanalysis (immunoblotting,
immunoprecipitation etc.), for functional activity tests of the
product, for studying proteolytic processing and so on. If a purification tag is fused to the cDNA of the protein, the synthesized
product can be directly purified from the in vitro translation
mix and used for subsequent studies. Commercially available in
vitro translation extracts also allow the co-translation of several
different proteins simultaneously, thus providing a system to
study protein:protein interactions as well as protein:DNA and
protein: RNA interactions.

Commonly exploited cell lysates from rabbit reticulocytes (Pelham and Jackson 1976) or wheat embryos (Shih and Kaesberg
1973) provide all the components necessary for translation
except mRNA. The latter allows the use of a defined template to
direct protein synthesis and avoid the high background resulting
from endogenous RNAs present in in vivo systems. Wheat germ
extract is also depleted from endogenous amino acids by chromatography; this system requires the addition of exogenous

**Principles
and
Applications**

[*] Corresponding author: Erkki Truve, phone: +372–6–398390;
fax: +372–6–398382; e-mail: erkki@kbfi.ee
[1] Institute of Chemical Physics & Biophysics, Akadeemia tee 23,
EE 0026 Tallinn, Estonia

amino acids. If radiolabeled amino acids are added to the reaction mixture, the lysate translates exogenous RNA to labeled protein(s), which can easily be identified by autoradiography following SDS-PAGE (see Chap. 7 by Sominskaya and Tars). In vitro translation is a good test for RNA quality, as only good quality full-length undegraded RNA can direct the synthesis of detectable amounts of the protein with the expected size.

The only requirement for template RNA, except for its purity and entity, is the presence of a translation initiation codon at the 5' proximity of the molecule and a translation termination signal at its 3' end. For optimal translation of the eukaryotic mRNA the AUG initiation codon must be in the Kozak consensus context (Kozak 1989) in case of rabbit reticulocyte lysate or in the conserved initiation context for higher plants (Lütcke et al. 1987) for wheat germ extract. However, noncanonical translation initiation at codons other than AUG can occur during in vitro translation, if the RNA template used is driving this type of protein synthesis initiation in vivo (Grünert and Jackson 1995; Schmitz et al. 1996). It has been shown that internal ribosomal entry sites can also be exploited by in vitro translation systems (Borman et al. 1995). When using in vitro transcribed RNA as the template, the presence of the 5'-end cap structure may enhance the activity of translation (Krieg and Melton 1984). However, we have noticed that in many cases capping of the RNA is not required for efficient product formation. Total RNA or a poly(A)+ RNA pool can also serve as an efficient template for the in vitro translation reaction. Provided that detection tools for a specific protein are available (e.g., specific antisera), this type of analysis can be used for the verification of the presence of a certain message in the RNA pool (Milanez and Mural 1989).

Materials

Equipment
- Thermostat for microcentrifuge tubes (Eppendorf Thermomixer 5436)
- UV spectrophotometer (Pharmacia Ultrospec 2000 UV/visible spectrophotometer)
- SDS-PAGE apparatus with DC power supply (Bio-Rad PROTEAN II xi System)
- Scintillation counter (Wallac 1414 WinSpectral liquid scintillation counter)
- Gel dryer (Hoefer vacuum gel dryer system)

- Autoradiography cassettes (Kodak BioMax cassette with Bio-Max MS intensifying screen)
- Film developing system (Kodak X-Omat M43A) or alternatively phosphoimager (Molecular Dynamics PhosphorImager SI)

Materials

- Glass fiber filters (Whatman GF/C)
- Whatman 3MM paper
- X-ray film (Hyperfilm βmax from Amersham)

Reagents

- Cell-free translation kit (Promega wheat germ extract, minus methionine; Promega rabbit reticulocyte lysate, nuclease treated) or cell-free coupled transcription/translation kit (Promega TnT T7/T3 coupled wheat germ extract system; Promega TnT T7/T3 coupled reticulocyte lysate system)
- [^{35}S]-labeled amino acid (Amersham L-[^{35}S]Methionine in vivo cell labeling grade)
- RNasin ribonuclease inhibitor (Promega Recombinant RNasin ribonuclease inhibitor)
- BMV RNA (Promega)
- Restriction endonucleases (from Fermentas, Pharmacia, Promega etc.)
- 3 M Na-acetate, pH 5.2
- Acetone
- Ethanol
- Klenow fragment of DNA polymerase I or T4 DNA polymerase (Promega)
- H$_2$O$_2$ (optional)
- Salicylic acid (optional)

Solutions and buffers

- Equilibrated phenol
  Melt the phenol at 68 °C.
  Add per 100 ml of phenol:
  0.1 g 8-hydroxyquinoline
  100 ml H$_2$O
  0.5 ml 10 M NaOH
  0.2 ml β-mercaptoethanol
  Mix and store at 4 °C.
- Chloroform/isoamylalcohol (24:1)
  Mix 24 vol of chloroform with 1 vol of isoamylalcohol. Store in closed bottle at room temperature.
- 100 % trichloroacetic acid (TCA) solution
  Dissolve 500 g of TCA in 227 ml of H$_2$O.

Note. The resulting solution will contain 100 % (w/v) TCA.

- Scintillation cocktail
  4 g 2,5-diphenyloxazole (POP)
  0.4 g 1,4-bis(5-phenyloxazol-2-yl)benzene (POPOP)
  1 l toluene
- Fixing solution for thin gels
  50 % methanol or 40 % ethanol
  10 % acetic acid
- Fixing solution for thick gels
  40 % ethanol
  7 % acetic acid
  5 % glycerol

It is important to avoid introducing RNase activity whenever possible. All equipment used during the in vitro translation procedure must not contain RNase. Microcentrifuge tubes and pipette tips should be autoclaved or otherwise treated to avoid RNase contamination (Sambrook et al. 1989). Wear gloves all the time for the same reason. Keep solutions used for RNA experiments separate from others.

8.1
Protein Synthesis In Vitro
in Wheat Germ Extract Using RNA Templates

This is the system of choice when template RNA (in vitro transcribed mRNA, viral genomic or subgenomic RNA, poly(A)+ RNA pool, etc.) is available. Generally speaking, the most commonly used eukaryotic cell-free translation systems – rabbit reticulocyte lysates and wheat germ extracts – are equally suitable for the analysis of most RNA species. Wheat germ extract is preferable for the translation of RNA preparations containing low concentrations of inhibitors to reticulocyte lysates, as well as for studying plant viral RNAs and several others, which for poorly understood reasons do not translate well in the reticulocyte system. It is a general strategy to choose the system which does not contain endogenously the product which is analyzed. For example, rabbit reticulocyte lysates contain much globin and several mammalian transcription factors. However, it has been shown that for translation of large RNA species the reticulocyte lysate system is more efficient than wheat germ extract. Here, we describe the experimental procedure for wheat germ extract, indicating the major differences between this system and rabbit reticulocyte lysates.

Procedure

1. Denature the template RNA by heating at 65 °C for 10 min and place *straight away on ice* for 3–5 min. Centrifuge briefly in microfuge to collect sample to the bottom of the tube.

Denaturation of Template RNA

Note 1. For each in vitro translation experiment, approximately 1 µg of template RNA is necessary.

Note 2. Placing of the samples quickly on ice is required for "snap cooling," to prevent renaturation of the RNA.

2. Prepare the following reaction mixture *on ice* in the following order:
   wheat germ extract 15.0 µl
   1 mM amino acid mixture minus methionine 2.4 µl
   1 M potassium acetate 2.0 µl
   RNasin ribonuclease inhibitor (40 U/µl) 0.6 µl
   RNA template 1.0 µg
   ^{35}S-methionine 370 MBq/ml, 10 mCi/ml 1.0 µl
   Nuclease-free dH$_2$O to total volume 30.0 µl

Preparation of the Reaction Mixture

Note 1. We are using wheat germ extract from Promega (cat. no. L4380). Equivalent extract can be purchased for example from Amersham. Store the wheat germ extract at −70 °C. *Do not freeze/thaw wheat germ extract more than twice.* If not all the extract is used immediately, it is necessary to subdivide the unused material into clean tubes, refreeze and store at −70 °C.

Note 2. The optimal potassium ion concentration varies from 50–200 mM, depending on the mRNA used. It is necessary to determine the optimal concentration of potassium for a particular mRNA.

Note 3. RNasin ribonuclease inhibitor in translation mixture prevents the degradation of RNA.

Note 4. The optimal template concentration depends of the translational efficiency of used template RNA. It is necessary to determine the optimal RNA template concentration.

Note 5. We recommend the use of a negative control reaction without RNA template to determine the background incorporation. For positive control reaction use 1.0 µg of brome mosaic virus (BMV) RNA (0.5 mg/ml). The optimal potassium concentration for translation of BMV RNA is 130 mM. In vitro translation of BMV RNA revealed five major proteins: 109, 94, 35, 20, and 15 kDa (Ahlquist et al. 1984).

Note 6. We are using in vivo cell labeling grade ^{35}S-methionine from Amersham (cat. no. SJ 1015) or translation grade ^{35}S-methionine from NEN (cat. no. NEG-009T). It is important to *store ^{35}S-methionine at −70 °C*, as otherwise it is easily oxidized to translation-inhibiting compounds.

Note 7. If more than one reaction is carried out, the preparation of premixes saves time and reduces the nuclease contamination.

Note 8. There are slight differences in the composition of the reaction mixture while using rabbit reticulocyte lysate. Usually we take 20 µl of reticulocyte lysate per 30 µl reaction volume. However, the reticulocyte lysate can also be diluted to 50 % without a substantial reduction in the reaction efficiency. As rabbit reticulocyte lysate contains endogenous amino acids, it is enough to take 0.5–1 µl of 1 mM amino acid mixture per 30 µl. The reaction can be carried out also without addition of exogenous amino acids. There is no need to add potassium ions to the rabbit reticulocyte lysate reaction.

In Vitro Translation Reaction

3. Mix gently by pipetting the reaction mixture up and down and collect the mixture by brief centrifugation in microfuge.

4. Incubate the reaction mixtures at 25 °C for 60 min.

Note. The temperature optimum for rabbit reticulocyte lysate is 30 °C.

5. Analyze the results of in vitro protein synthesis measuring the incorporation of radiolabel and/or on standard SDS polyacrylamide gels as described by Laemmli (1970; see Chap. 7 by Sominskaya and Tars).

8.2
Protein Synthesis In Vitro Using DNA Templates and TNT T7 Coupled Wheat Germ Extract System

A coupled in vitro transcription/translation system is the right choice when the translation is driven by a DNA template. The use of the coupled system eliminates the need to prepare RNA separately prior to the conventional in vitro translation procedure. Interestingly, it has been shown that coupled transcription/translation also yields more protein than standard cell-free translation (Thompson et al. 1992). Template DNA must, in addition to the necessary translation signals, also contain the appropriate RNA polymerase promoter sequence. Coupled systems for bacteriophage T7, T3, or SP6 RNA polymerases are available. The suitable one for each particular experiment is determined by the presence of the corresponding promoter in the DNA construct. For T3 and SP6 systems best results have been obtained with circular DNA templates. However, when T7 RNA polymerase is used, it is recommended to add linearized plasmid into the reaction mix. The addition of the cap analog to the transcription/translation reaction does not enhance the expression levels (Thompson 1993). It should be added that in case the needed signals are included into the molecule, PCR products can also serve as the template DNA for a coupled transcription/translation reaction. It is possible to express multiple genes simultaneously using this system, even if they are under the control of different phage polymerase promoters (provided that all needed RNA polymerases are added to the reaction mix).

▪ Procedure

1. Linearize the plasmid template DNA with a unique restriction enzyme at the site located at the 3' end of the DNA insert, opposite from the T7 RNA polymerase promoter.

    Linearization of Plasmid DNA

Note 1. Plasmid DNA isolated from CsCl density gradients is sufficiently pure for this experiment. You can also use commercially available systems for plasmid DNA purification (for example Promega Wizard Midipreps DNA purification system, QIAGEN Plasmid Midi Kit).

Note 2. For each in vitro transcription/translation experiment, 1 µg of linearized template DNA is necessary. We generally linearize enough template DNA for five reactions.

Note 3. For plasmid linearization it is important to avoid the use of restriction enzymes which produce 3' overhanging ends. If no alternative is available, the linearized template 3' overhanging ends can be blunted with the Klenow fragment of *E. coli* DNA polymerase I or with T4 DNA polymerase (Sambrook et al. 1989).

Note 4. Complete restriction enzyme digestion of the plasmid is essential because the end of the linear template acts as the termination site for transcription. A small aliquot of the digested template must be checked on agarose gel electrophoresis (see "Support Protocol" in Chap. 6 by Pata and Truve).

Note 5. It has been shown that if DNA constructs under the control of T3 or SP6 RNA polymerase promoters are used, it is not necessary to linearize the template DNA prior to use (Thompson and Van Oosbree 1992).

2. Extract the sample twice with 1 vol phenol to remove proteins. Extract the aqueous phase once with 1 vol chloroform/isoamyl alcohol (24:1).

3. Precipitate the DNA using the standard procedures with 1/10 vol 3 M sodium acetate and 2.5 vol 96 % ethanol. Wash the pellet once with cold ($-20\,°C$) 70 % ethanol and dry the pellet.

4. Resuspend the DNA in RNase-free water, take 1 µl aliquot and measure the OD at 260 nm. The DNA is now ready to use.

Note. The optimal concentration of DNA in the sample is 0.5–1.0 mg/ml.

Additional note: Preparation of the linearized DNA template will take approximately 5 h or more, depending on the time required for the restriction digestion.

Preparation of the Reaction Mixture
5. Prepare the following reaction mixture *on ice* in the following order:
TnT wheat germ extract 12.5 µl
TnT reaction buffer 1.0 µl
1 mM amino acid mixture minus methionine 0.5 µl
RNasin ribonuclease inhibitor (40 U/µl) 0.5 µl
DNA template(s) in nuclease-free dH$_2$O 9.0 µl

^{35}S-methionine 370 MBq/ml, 10 mCi/ml 1.0 µl
TnT T7 RNA polymerase 0.5 µl

Note 1. We are using the TnT T7 coupled wheat germ extract system obtained from Promega (cat. no. L4140). Store the components of the kit at −70 °C. *Do not freeze/thaw wheat germ extract more than twice.*

Note 2. RNasin ribonuclease inhibitor in translation mixture prevents the degradation of RNA.

Note 3. Use 0.5–1.0 µg of DNA template in one reaction mixture. This system also allows the co-translation of more than one gene product in the same reaction by using two different DNA templates. The total amount of added DNA templates must be less than 1.0 µg.

Note 4. We recommend using a negative control reaction without DNA template to determine the background incorporation. For positive control reaction use 1.0 µl of luciferase control DNA (Promega, 0.5 mg/ml). Luciferase is a monomeric protein of approximately 61 kDa molecular mass (Thompson 1993).

Note 5. We are using in vivo cell labeling grade ^{35}S-methionine from Amersham (cat. no. SJ 1015) or translation grade ^{35}S-methionine from NEN (cat. no. NEG-009T). It is important to *store ^{35}S-methionine at −70 °C.*

Note 6. If more than one reaction is carried out, the preparation of premixes saves time and reduces the nuclease contamination.

Note 7. This protocol is also suitable using TnT Coupled Reticulocyte Lysate System from Promega (cat. no. L4610).

6. Mix gently by pipetting the reaction mixture up and down and collect the mixture by brief centrifugation in microfuge.

7. Incubate the reaction mixtures at 30 °C for 90 min.

8. Analyze the results of in vitro protein synthesis measuring the incorporation of radiolabel and/or on standard SDS polyacrylamide gels as described by Laemmli (1970; see Chap. 7 by Sominskaya and Tars).

Coupled in Vitro Transcription/ Translation Reaction

8.3
TCA Precipitation – Spotting

Precipitation of the synthesis products with trichloroacetic acid (TCA) enables rapid evaluation of the efficiency of the in vitro translation reaction. The negative control reaction (where no RNA was added to the reaction mix) serves here for the determination of the non-specific background incorporation of labeled amino acids. Usually mRNA preparations give a stimulation over background of about 10- to 50-fold.

▨ Procedure

Applying the Sample to the Filter

1. To measure the amount of incorporation of a radiolabeled precursor into the proteins, apply a known volume of each sample (normally 2.5 to 5.0 µl) to the center of a dry glass fiber filter.

Note 1. We are using Whatman GF/C glass fiber filters (∅25 mm).

Note 2. The filters should be marked prior to use.

2. Allow the filters to dry at room temperature.

Washing the Filters

3. Place the filters into the glass flask and add the ice-cold 5 % TCA. Incubate on ice for 5 min.

4. Incubate the filters for additional 10 min at 90 °C.

5. Drain the TCA and add fresh ice-cold 5 % TCA. Incubate on ice for 5 min.

6. Rinse the filters thoroughly four or five times with tap water, once with 95 % ethanol and once with acetone.

Note. If the rabbit reticulocyte lysate system was used for protein in vitro translation the additional treatment with 10 % H_2O_2 is necessary before the washing steps.

7. Allow the filters to dry at room temperature.

Radioactivity Counting

8. To determine the incorporation of radiolabel, add the dry filters to the counting vials, add the scintillation mixture, and count.

Note. Several commercially available pre-made scintillation cocktails can be used.

8.4
Fixing and Drying of SDS-Polyacrylamide Gels

This protocol is suitable for the fixation and drying of polyacrylamide gels during different applications. Fixing of the gel prior to drying is needed to avoid gel cracking. Dried gels are much better for sensitive autoradiography as water is a very efficient protective screen for radioactive isotopes.

Procedure

1. Remove the gel from the electrophoresis apparatus and mark the orientation by cutting off one corner.

    Fixing the Gel

2. Fix the gel at room temperature in 5–10 gel volumes of fixing solution for 30–60 min to dehydrate and shrink the gel.

Note 1. For gels ≤0.75 mm thick the fixing solution contains 50 % methanol and 10 % acetic acid.

Note 2. For gels thicker than 0.75 mm add to the above mentioned fixing solution 3–5 % glycerol. The glycerol in the fixing solution will minimize the cracking of the gel.

Note 3. We also use a fixing solution containing 40 % ethanol, 7 % acetic acid and 5 % glycerol.

3. Place the gel to the piece of Whatman 3MM paper before drying.

    Drying the Gel

Note. Drying the gels on Whatman 3MM paper minimizes the shrinkage and distortion of the gel.

4. Dry the gel with a vacuum gel dryer according to the dryer instructions.

Note. The drying time depends on percent of acrylamide and gel thickness. Typically thin gels dry in under 2 h.

Results

Figure 8.1 represents the incorporation rate of labeled amino acids into the in vitro translated products and its dependence on the RNA concentration. Note that the translation efficiency for cocksfoot mottle virus (CfMV) RNA increases only until the RNA reaches the concentration of 3.75 µg/30 µl. At this point the reaction will be saturated and further increase in RNA concentration does not increase the reaction efficiency but rather decreases it. The saturation point will be different for different RNA templates used, but the general effect will remain the same.

Figure 8.2 demonstrates the effect of the duration of incubation on the synthesis of in vitro translation products. The overall amount of products synthesized increases throughout 50 min, which was the total length of this experiment. The time optimum is slightly different for different RNAs, but generally it can be concluded that 60–90 min is needed to obtain the highest concentrations of reaction products. Note that at least in the case of

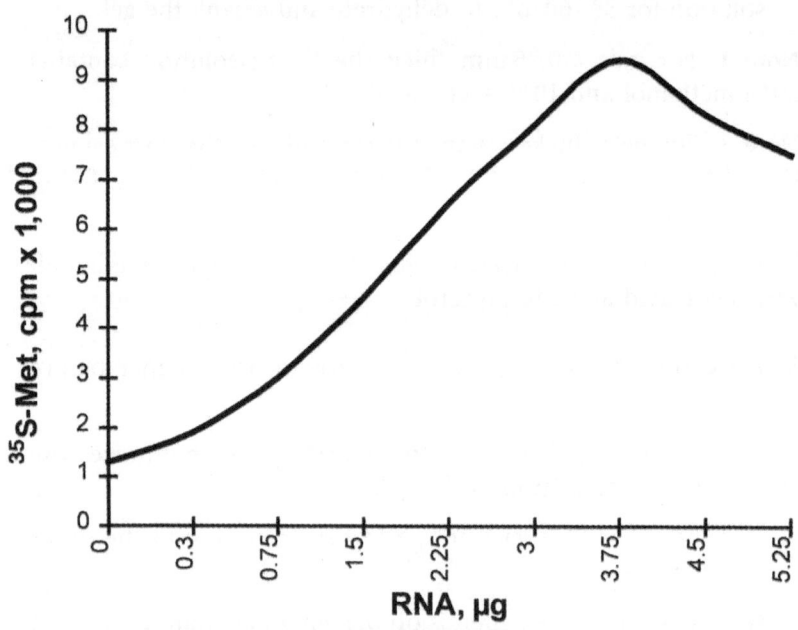

Fig. 8.1. Dependence of in vitro translation efficiency on CfMV RNA concentration. Cell-free protein synthesis was performed in wheat germ extract with ^{35}S-methionine. The activity of in vitro translation was assessed by precipitating the synthesis products with TCA and determining the amount of incorporated radioactivity in liquid scintillation counter

Fig. 8.2. Time course of CfMV RNA translation in wheat germ extract. 1 µg of RNA was translated. The ^{35}S-methionine-labeled translation products were resolved on 12.5 % PAAG-SDS and detected by autoradiography for 12 h. To the *right* the molecular mass markers (kDa), at the *top* the analyzed time points

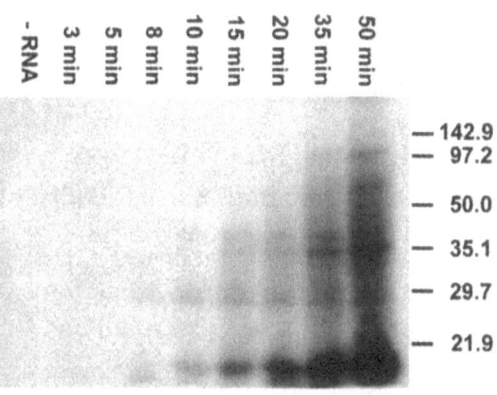

CfMV detectable amounts of smaller in vitro translation products are produced earlier than larger polypeptides. Note also that by the end of the incubation, in addition to the full-length polypeptides, many degradation and partial synthesis products appear. This means that to obtain pictures with low background levels it is sometimes necessary to shorten the incubation period.

Coupled in vitro transcription/translation of CfMV coat protein cDNA, demonstrated in Fig. 8.3, shows that as little as 20 ng of template DNA is sufficient to produce a clearly visible reaction product when the dried SDS-PAAG is autoradiographed. Depending on DNA, the reaction efficiency reaches a plateau at about 0.5–1 µg of DNA per reaction. When exposure to X-ray film is prolonged, even less than 20 ng DNA can produce detectable amounts of protein.

Fig. 8.3. Expression of the coat protein gene of CfMV using the TNT wheat germ system. Different concentrations of linearized DNA construct were used for the T7 TNT reaction (the concentrations used are shown at the *top*). The translation products were separated on 12.5 % PAAG-SDS and detected by autoradiography for 2 h

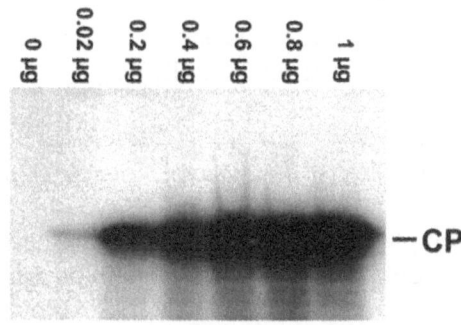

Troubleshooting

Low Translation Efficiency

- The reaction mixture has been contaminated with nucleases.
  - Add RNasin ribonuclease inhibitor to the reaction mixture (if it is not already added).
  - Avoid contamination of the reaction mixture with RNases while creating the mixture (wear gloves, use RNase-free microcentrifuge tubes and pipet tips, etc.).

- Template RNA/DNA translational efficiency is low.
  - Optimise template concentration for translation mixture by serially diluting your template first and then adding the same volume of the template to each reaction.
  - Verify that template RNA/DNA is of high quality and does not contain nucleases.

- The radioisotope has been decayed.
  - Use functional radioisotope. Remember that the half-life for ^{35}S is 87.4 days.

Low RNA-Directed Translation Efficiency

- Template RNA contains inhibitors of translation (e.g., double-stranded RNA, polysaccharides).
  - Co-translate template RNA with control RNA and perform control translation using only control RNA. Evaluate the efficiencies of both translations. In case of higher efficiency of control translation, template RNA contains translational inhibitors.

- The concentration of potassium and/or magnesium ions is not optimal.
  - Optimize the potassium and/or magnesium ions concentration for the template RNA.

Low DNA-Directed Translation Efficiency

- Template DNA is not completely linearized.
  - Control the DNA linearization on agarose gel electrophoresis.

- Template DNA contains RNases.
  - DNA digestion must be followed by phenol extraction and ethanol precipitation of the template.

No Translation

- The cell-free extract/lysate has been decomposed.
  - Use functional cell-free extract/lysate and test it before using in experiment. Avoid freezing/thawing the extract/lysate more than twice.
- Added template does not work.
  - Using RNA templates, verify that RNA has not degraded during the purification or in vitro transcription.
  - Using DNA templates, verify that your construct contains required RNA polymerase promoter and translational initiation codon.
- The radioisotope has been decayed.
  - Use fresh, functional radioisotope.
- RNA polymerase has been inactivated for coupled synthesis.
  - Order a new system.

Excessive Diffuse Signal

- Overload of protein.
  - Apply less protein to the gel.
- Improper gel conditions.
  - Optimize gel and electrophoresis conditions.
  - Increase acrylamide concentration of the gel.
  - Check gel and buffers recipes and quality.
  - Start the run immediately after loading to prevent diffusion.
  - Reduce the run speed by decreasing the current strength (normally 20 mM per gel).

High Background

- Many probable causes
  - After electrophoresis, soak the gel in H_2O for 15–30 min prior to drying.
  - Use salicylic acid treatment: soak the gel after fixing in H_2O for 30 min and then in 16 % salicylic acid prior to drying.
  - After the completion of the reaction, treat your in vitro translation mix with RNase A (at final concentration of 0.2 mg/ml) for 5 min. This avoids the background labeling of aminoacyl tRNAs.

Gel Cracking During Drying

- Many probable causes
  - Add glycerol to the fixing solution prior to drying.
  - Do not remove the vacuum before the gel is dry.
  - Be sure that the vacuum created for drying is sufficient; gel cracking may occur due to insufficient negative pressure.

References

Ahlquist P, Dasgupta R, Kaesberg P (1984) Nucleotide sequence of the brome mosaic virus genome and its implications for viral replication. J Mol Biol 172:369–383

Borman AM, Bailly JL, Girard M, Kean KM (1995) Picornavirus internal ribosome entry segments: comparison of translation efficiency and the requirements for optimal internal initiation of translation in vitro. Nucleic Acids Res 23:3656–3663

Grünert S, Jackson RJ (1995) The immediate downstream codon strongly influences the efficiency of utilization of eukaryotic translation initiation codons. EMBO J 13:3618–3630

Kozak M (1989) The scanning model for translation: an update. J Cell Biol 108:229–241

Krieg P, Melton, D (1984) Functional messenger RNAs are produced by SP6 in vitro transcription of cloned cDNAs. Nucleic Acids Res 12:7057–7070

Laemmli UK (1970) Cleavage of structure proteins during assembly of the head of bacteriophage T4. Nature 227:680–685

Lütcke HA, Chow, K, Mickel FS, Moss KA, Kern HF, Scheele GA (1987) Selection of AUG codons differs in plants and animals. EMBO J 6:43–48

Milanez S, Mural RJ (1989) Cell-free translation of plant mRNA in rabbit reticulocyte lysate. Promega Notes 19:3

Pelham HRB, Jackson, RJ (1976) An efficient mRNA-dependent translation for reticulocyte lysate. Eur J Biochem 67:247–256

Sambrook J, Fritsch EF, Maniatis T (1989). Molecular cloning. A laboratory manual, 2nd edn. Cold Spring Harbor Laboratory, New York

Schmitz J, Prüfer D, Rohde W, Tacke E (1996) Noncanonical translation mechanisms in plants: efficient in vitro and in planta initiation at AUU codons of the tobacco mosaic virus enhancer sequence. Nucleic Acids Res 24:257–263

Shih D, Kaesberg, P (1973) Translation of brome mosaic viral ribonucleic acid in a cell-free system derived from wheat embryo. Proc Natl Acad Sci USA 70:1799–1803

Thompson D (1993) Coupled in vitro transcription/translation using TnT^{TM} wheat germ extract. Promega Notes 40:1–3

Thompson D, Van Oosbree T (1992) TnT^{TM} lysate coupled transcription/translation: comparison of the T3, T7 and SP6 systems. Promega Notes 38:13–14

Thompson D, Van Oosbree T, Beckler G, Van Herwynen J (1992) The TnT^{TM} lysate systems: one step transcription/translation in rabbit reticulocyte lysate. Promega Notes 35:1–4

Suppliers

Amersham International plc
Little Chalfont
Buckinghamshire, HP7 9NA
ENGLAND
http://www.amersham.co.uk

Bio-Rad Laboratories
2000 Alfred Nobel Drive
Hercules, California 94547
USA
http://www.bio-rad.com

Eastman Kodak Company
343 State Street
Rochester, New York 14650–1139
USA
http://www.kodak.com

Eppendorf-Netheler-Hinz GmbH
D-22331 Hamburg
GERMANY
http://www.eppendorf.com

Fermentas AB
V. Graiciuno 8
LT-2028 Vilnius
LITHUANIA
http://www.fermentas.com

Hoefer Pharmacia Biotech Inc.
654 Minnesota Street
San Francisco, California 94107-0387
USA
http://www.hpb.com

Molecular Dynamics
928 East Arques Avenue
Sunnyvale, California 94086-4520
USA
http://www.mdyn.com

NEN Life Science Products
549-3 Albany Street
Boston, Massachusetts 02118
USA
http://www.nenlifesci.com

Pharmacia Biotech AB
S-751 82 Uppsala
SWEDEN
http://www.biotech.pharmacia.se

Promega Corporation
2800 Woods Hollow Road
Madison, Wisconsin 53711-5399
USA
http://www.promega.com

QIAGEN GmbH
Max-Volmer-Strasse 4
D-40724 Hilden
GERMANY
http://www.qiagen.com

Wallac Oy
P. O. Box 10
FIN-20101 Turku
FINLAND
http://www.wallac.fi

Whatman International Ltd.
Whatman House
St. Leonard's Road
20/20 Maidstone
Kent ME 16 OLS
ENGLAND
http://www.whatman.co.uk

Subject Index